Remote Sensing Technology in Forensic Investigations

Remote Sensing Technology in Forensic Investigations

Geophysical Techniques to Locate Clandestine Graves and Hidden Evidence

G. Clark Davenport

CRC Press
Taylor & Francis Group
Boca Raton London New York

CRC Press is an imprint of the
Taylor & Francis Group, an **informa** business

CRC Press
Taylor & Francis Group
6000 Broken Sound Parkway NW, Suite 300
Boca Raton, FL 33487–2742

CRC Press is an imprint of Taylor & Francis Group, an Informa business

Printed on acid-free paper

International Standard Book Number-13: 978-1-138-73297-1 (Hardback)
International Standard Book Number-13: 978-1-315-18657-3 (eBook)

Visit the Taylor & Francis Web site at
http://www.taylorandfrancis.com

and the CRC Press Web site at
http://www.crcpress.com

Printed and bound by CPI Group (UK) Ltd, Croydon, CR0 4YY

Contents

List of Figures and Tables

Figures

Tables

Foreword

One of Denver's more notorious serial killers buried two female victims in his backyard, just a few steps from his backdoor. This killer lived in my police district, and as a consequence, I became acquainted with Clark Davenport and NecroSearch. Clark got my attention. He was not a cop, but could have been mistaken for one. There he was in a major homicide crime scene dragging this big red metal box around the killer's backyard, periodically stopping to insert little red flags in the ground. I had not witnessed this behavior before at other crime scenes during my 35 year career as a police. When I decided to disturb him, Clark informed me that he was "identifying disturbances in the ground" not, as he has often reminded me, and everyone else who asks, "finding bodies"! Clark is a natural teacher and on that day he was very agreeable, in spite of me interfering in his work, to explain what he was doing and why. This meeting, by the way is, in part, why Clark is a professor at Regis University today. The advantages of this big red "iron contraption" for law enforcement, and especially for homicide investigators, he disclosed, were critical for identifying buried criminal evidence. What Clark had to say that afternoon is comprehensively provided for the reader in this text, a long-overdue analysis addressing what is frequently a little understood subject in criminalistics: finding clandestine graves. It is important to understand its application for both crime scene investigators and criminal justice students who hope to enter the criminology field.

To provide a taste of what you can expect in this well-informed book is that "dragging a metal box around," in fact a ground-penetrating radar (GPR), provides the crime scene investigator with Superman's x-ray Vision, of sorts. And while Clark does not like to talk about locating bodies and rather prefers demonstrating what it does provide—finding disturbances in the ground—I know that the lay person is hard pressed not to believe a "body" absolutely cannot be found in this manner.

From that afternoon until now, several years later, Clark and I have formed a professional and cordial relationship that has benefited hundreds of criminology students, and hopefully will benefit those in law enforcement, who will now be aware of this important scientific investigatory tool. Every course I teach at Regis University includes, by one means or another, crime scene investigation. Clark has provided my students with a strong background in the science of remote sensing, and its use in finding clandestine evidence; I fear that few students, and not many criminal investigators, are aware of the different methodologies. In this important criminalistics text providing the value of GPR, Clark has afforded the reader with the importance of this investigatory tool. If you have not heard about GPR and other geophysical techniques, and their applications for identifying clandestine evidence, in many cases, clandestine graves, this text is a must read for you, and as a refresher for those who seek updated material. It belongs to every homicide investigator's book shelf.

Don E. Lindley, Ph.D.
Associate Professor of Criminology
Regis University

A Caution

I have spent the last 29 years applying remote sensing techniques in forensic investigations. That these techniques save time, manpower, and money is readily apparent when a vehicle containing a murder victim can be found in the Missouri River in less than 2 days (having been there 7 years) or the location of a murder victim can be found under a concrete slab (after 28 years) in a little over 4 hours.

I have one caution, however. You, the criminal investigator, using tenacity, skill, and an understanding of human nature, have been the common factor in all the successful investigations in which remote sensing has played a role. Please, do not rely on high-technology gadgets and wizardry at the expense of your skills and dedication!

This publication is intended for one purpose, and one purpose only—to help you work with and successfully manage personnel, instrumentation, and materials to find hidden evidence and buried bodies.

While this publication may suggest actions to meet standards expected of members of law enforcement, it is not and cannot be exhaustive of all necessary steps to take, and is not meant to be in any manner a substitute for protocol or guidelines of any law enforcement organization. In other words, do not rely solely on this manual for guidance.

Similarly, any discussion of legal and operational issues in this publication is intended to provide an overview only, and not to provide any legal opinions. You should seek counsel from your own professional advisors as to any such matters.

Acknowledgments

I wish to sincerely thank the officers and members of NecroSearch International, for their support and assistance in developing the use of geophysical surveys in forensic investigations. In particular, special thanks is due to Mr. Jim Reed, director of Research and Development, RockWare Incorporated, and Mr. Timothy Deignan, senior geophysicist, HydroGeoLogic, Inc. The author is solely responsible for any errors and omissions in this publication.

"I shall be as secret as the grave"

—Miguel de Cervantes, Don Quixote

About the Author

Clark Davenport's career spans more than 50 years of planning, managing, and performing remote sensing surveys for criminal, environmental, groundwater, geotechnical, mining, and archaeological investigations on six continents. He is a cofounder of NecroSearch International and the founder of GeoForensics International. Since 1987, he has pioneered the use of remote sensing in criminal investigations and has personally assisted in more than 85 criminal investigations in 32 states and 7 countries. He serves as a consultant to municipal, federal, and foreign law enforcement agencies including the FBI, U.S. Secret Service, Drug Enforcement Administration, Bureau of Alcohol Tobacco and Firearms, Spanish Guardia Civil, Mexican, Guatemalan Forensic Foundation, Federal Court of Colombia, Australian New South Wales Police, and the Russian FSB.

He currently works as senior environmental engineer (a contract position) within the Underground Injection Control Enforcement Group, U.S. EPA Region 8, and teaches forensic science at Regis University. He is a past instructor at the Colorado Law Enforcement Training Academy, the Community College of Denver, and Metropolitan State College of Denver and invited special topics instructor at the FBI Academy.

Davenport is a decorated Vietnam combat veteran. He has a degree in geophysical engineering from the Colorado School of Mines (1964) and an associate degree in criminal justice from Red Rocks Community College (1997). Davenport is a California registered geophysicist and is professionally proficient in Spanish.

1 Introduction

Applying geophysical methods to locate physical evidence is by no means the ultimate panacea for the criminal investigator. Geophysical methods are based on detecting and measuring physical, electrical, and chemical contrasts within the earth. Contrasts between the physical evidence being sought and the materials within the earth may be very small and require the use of sensitive equipment to detect. The success of a geophysical investigation is a function of numerous factors including site conditions, the physical characteristics of the evidence being sought, and the experience of the individuals conducting the geophysical surveys.

Geophysical surveys are nondestructive because only indirect measurements are made. In general, geophysical measurements do not require excavations, probing, or drilling. Recent developments in geophysical instrumentation, locating equipment, and computer technology have increased the successful application of geophysical surveys in near-surface (0–10 m) investigations, depths at which many perpetrators work within while attempting to dispose of bodies or evidence clandestinely.

1.1 Historical Background

The foundations of geophysics were developed several hundred years ago as the understanding of the natural sciences developed. As early as the 1600s, magnetic compasses were employed to locate ferrous iron deposits in Sweden. The development of modern geophysical techniques for mining and petroleum exploration began during the period of 1910–1930. The applications of geophysical techniques for engineering and construction projects date to the 1920s. The applications of geophysical techniques for archaeological investigations date to the

1940s, for forensic investigations to the 1970s, and for medical purposes to the 1980s.

Major advancements in geophysical technology have come about as the results of military applications. Acoustic techniques were used during both World Wars to locate enemy artillery batteries, while magnetometers and sonar systems were utilized during World War II to search for mines and submarines. Seismic detectors were used during the Vietnam War to provide indirect information on enemy movements. Seismic and ground-penetrating radar (GPR) techniques were used for tunnel detection during the Vietnam War and to detect tunnels running under the demilitarized zone between North and South Korea. Electromagnetic (EM) techniques are being refined and used to map ice thickness in polar regions, information of vital importance to submariners.

Although there are numerous geophysical surveying methods, magnetic (MAG), EM, and GPR techniques are the primary ones successfully used in archaeological investigations. Owing to those successes, those surveying methods have been incorporated into forensic investigations.

1.2 General Methodology

Geophysical land methods require the movement of an instrument package over a site, normally along preestablished lines. In some surveys using older instruments, the instrument operator stops at predetermined stations along each line and obtains a measured value. Once the measured value is recorded, the operator advances to the next station. Newer instruments allow for the continuous collection of data. Detailed discussions of data acquisition procedures for each technique are present in the following sections.

Geophysical marine and airborne methods also involve the continuous collection of data, however, not normally along predetermined lines or flight paths. One of the major difficulties with marine and airborne surveys is determining the location of the instrument packages.

1.3 Need for High-Resolution Geophysical Capabilities

The detail necessary to search for physical evidence or a clandestine grave requires the use of high-resolution geophysical techniques. For example, a perpetrator may be frightened, lazy, and in a hurry and thus not go to great lengths (or depths) to dispose of a body; the burial may only produce small geophysical contrasts between the disturbed grave system and the surrounding undisturbed soils. High-sensitivity geophysical instrumentation and precise field methods are required to evaluate these small, near-surface contrasts. This means that geophysical surveys have to be planned to take measurements across a site, while obtaining high-quality data that do not contain signals from outside influences (e.g., noise from nearby features or from the instrument itself).

1.4 Site Survey Grid

Prior to discussing geophysical survey methods and field procedures, it is necessary to understand how a particular site will be laid out in a grid system, with survey lines, such that the results of any geophysical survey and any further work (excavation of evidence and/or human remains) may be directly related to the maps of the site being investigated. The use of a survey grid is critically important to establish spatial relationships of any evidence detected by geophysical surveys and recovered in an excavation process. Spatial relationships are important to the criminal investigator when viewing evidence and interpreting a sequence of events.

In forensic geophysical surveys, the target, or evidence, is often small compared to the size of the area being investigated. In general, the grid lines and station spacing along the grid lines should be spaced no further apart than 1–2 times the size of the target. For example, using GPR to locate anomalies that may relate to the disposal of an infant will require closer line and station spacing than using GPR to locate anomalies that may relate to the disposal of an adult.

2 Overview of Modern Geophysical Methods

Geophysical methods are based on the ability to detect and measure contrasts between the characteristic properties of materials. All materials exhibit some distinguishing characteristic properties. Contrasts exist owing to differences in density, electrical (conductive) properties, magnetic susceptibilities, and chemical and mineral properties. Geophysical methods provide a means to investigate these properties actively or passively. When materials are buried in the subsurface, the characteristic properties may be masked by the properties of the encapsulating materials, thereby making any contrasts between the buried and encapsulating materials indistinguishable. If a contrast is too small to be measured or does not exist, geophysical methods are useless.

2.1 Passive Geophysical Methods

Passive geophysical measurements are based on measuring contrasts within the earth that exist owing to naturally occurring fields created by earth-related processes. For example, naturally occurring electric currents (telluric currents) in the core and within the mantle of the earth create a natural magnetic field that can be measured at any point on the earth's surface. Water flowing within the subsurface generates a minute electrical field that can be measured.

Passive geophysical methods do not require the introduction of a signal (energy) into the earth to create a measurable response. Instrumentation used for passive geophysical methods requires only a receiver to detect the signals from the earth's natural processes.

The four classical passive geophysical methods are as follows:

- Magnetics
- Self-potential

- Gravity
- Thermal imaging

Of these, only two, magnetics (MAG) and thermal imaging, are primarily applicable to forensic investigations, in the author's opinion. Magnetic surveying techniques are discussed in detail in Chapter 4, and thermal imaging surveying techniques are discussed in Chapter 8.

2.1.1 Magnetics

Utilizing sensitive instruments, the earth's total magnetic field can be measured with great precision, and contrasts within the total magnetic field can be delineated. The contrasts are owing to a phenomenon called magnetic susceptibility.

The magnetic susceptibility, or the ease at which materials can be magnetized, is the basis for this technique. *Ferrous material* is any material that can rust and be magnetized. Local disturbances, or anomalies, measured within the earth's total magnetic field can indicate the position of buried ferrous objects and even displaced soil materials. Heated objects that contain even small amounts of magnetic material exhibit a high magnetic response. For example, magnetic methods are very valuable in archaeological investigations for detecting fire hearths.

2.1.2 Self-Potential

Self-potential (SP) methods provide a means of measuring electrical anomalies created by the flow of fluids, heat, and ions in the earth. SP method has been extensively utilized in geothermal exploration and to detect leaks in dams, canals, and reservoirs. As opposed to electrical resistivity surveying, no active electrical current is introduced into the ground to create a measurable response. SP surveys have been utilized, on an applied research basis, in archaeological and forensic investigations; however, most of the results have been ambiguous.

2.1.3 Gravity

Measurements of the earth's gravitational field with accurate precision have been used to map subsurface geological structure based on localized and regional rock density contrasts. Gravity surveys are very useful in mapping faults, buried channels, and salt domes. Microgravity surveys have been used in engineering investigations to map subsurface cavities, and in archaeological investigations to locate catacombs and voids within large structures. The use of microgravity surveys in forensic investigations is very limited owing to the extreme measurement precision necessary and the time-consuming fieldwork required.

2.1.4 Thermal Imaging

Measurements of the differences (contrasts) in reflected heat can provide useful information for the criminal investigator. Near-surface disturbances in the ground will heat up differently from the surrounding ground during daytime.

In the evening, as materials start to cool, near-surface disturbances will reflect their stored heat differently than the surrounding ground. This contrast in reflected heat can be imaged by special cameras and then digitally processed. Thermal imaging can be used, for example, to detect anomalies behind walls and under floors. The author performed a thermal imaging survey in the basement of a home in an effort to determine if a hidden room existed behind the concrete walls. Once all furnishing and wall covering were removed from the basement, a large heater was used to heat the room. Once a maximum heat was reached, the heater was turned off, and the concrete walls were scanned, at time intervals, looking for anomalous areas.

2.2 Active Geophysical Methods

Active methods involve the introduction of a signal (physical, electrical, electromagnetic, or acoustic) into the subsurface. The interaction of these signals with subsurface materials of contrasting properties produces a return signal (response), which can be measured by appropriate geophysical instruments. All active geophysical methods require instrumentation that includes a transmitter to transmit energy into the subsurface and a receiver to detect the response from subsurface materials.

The classical active geophysical methods include the following:

- Electromagnetics (EM)
- Ground-penetrating radar (GPR)
- Electrical resistivity
- Acoustic (seismic)

Active geophysical methods primarily applicable in forensic investigations are EM and GPR. Details of EM techniques are presented in Chapter 5, and details of GPR techniques are presented in Chapter 6.

2.2.1 *Electromagnetics*

Electromagnetic surveying makes use of time-varying, low-frequency (a few hundred to one thousand cycles per second) electromagnetic fields induced into the earth. The induction of an electromagnetic current into the earth results in a primary electromagnetic field being formed. When the primary field comes into contact with metallic objects within the subsurface, a secondary field results. Measurements of the relationship between the primary and secondary fields are used to determine the characteristics of the metallic object. EM surveys play a major role in groundwater exploration, in mapping salt water intrusion, and in unexploded ordnance mapping. EM surveys are also very useful in archaeological and forensic investigations.

EM surveys can be used to detect both *ferrous* and *nonferrous* (materials not capable of being magnetized) metals and also to detect changes in soil conductivity. Conductivity changes in soils may be related to changes in moisture or chemical content. Many metal detectors operate using electromagnetic fields induced into the subsurface. EM surveys are used by many farmers to determine the moisture content of cropland and the need for fertilizer application. In forensic investigations, EM surveys play a vital role in mapping subsurface moisture content that can be used to delineate clandestine graves.

2.2.2 *Ground-Penetrating Radar*

GPR is based on the transmission of high-frequency (a few thousand to one million cycles per second) electromagnetic waves into the subsurface and recording the electromagnetic energy scattered back to the surface by reflecting objects. The reflections are caused by contrasting materials of different dielectric properties. The depth of investigation with GPR is generally quite shallow owing to the inherent attenuation of low-frequency electromagnetic waves by the earth. This disadvantage is partially offset by the increased resolution of radar compared to other geophysical methods.

GPR has effectively mapped soil layers, depth of bedrock, cavities, voids, rock fractures, ice thickness, and buried stream channels. In engineering and hazardous waste environmental studies, it has been utilized to locate and delineate areas of buried waste materials, contaminant plumes, and buried utilities and to examine concrete structures. Archaeological applications include examination of burial sites, buried structures, and the detection of metallic objects.

GPR surveys will not delineate human remains. GPR surveys will locate subsurface anomalies that might contain human remains; however, this will only be evident upon excavation of the anomalies. The author has used GPR surveys for forensic investigations for over 25 years; however, when asked by law enforcement personnel to use GPR, the author states that there is good news and bad news when using GPR. The good news is GPR will locate anomalies. The bad news is GPR will locate anomalies. Only when the results of a GPR survey (or any geophysical survey) are integrated with other site- and case-specific information will the probability that a GPR anomaly contains human remains be known. And only with excavation will the actual cause of the anomaly be known.

2.2.3 *Electrical Resistivity*

Electrical resistivity surveying is based upon the contrast in flow of an injected electrical current within the earth. The current flow is primarily dependent upon water saturation and salt content in the subsurface. This method measures the ease with which a transmitted electrical current will flow through the subsurface materials. Observed lateral and vertical contrasts in electrical resistivity across a site can be indicative of geological information. Resistivity surveying is used in mineral and groundwater exploration programs and in electrical grounding studies.

The results of electrical resistivity surveying are often comparable to electromagnetic surveying, because the resistivity of a material is the reciprocal of the conductivity of the same material, and electromagnetic surveying gives results in terms of conductivity. However EM surveying is much less labor intensive and faster than resistivity surveying, and therefore EM surveying is more adaptable to forensic investigations.

2.2.4 *Acoustic (Seismic and Sonar)*

Acoustic methods such as seismic or sonar are based on the velocity at which acoustic waves travel through materials of different densities. The method involves transmitting acoustic energy into the earth and measuring the time it takes the signal to travel from the transmitter (source) to a receiver implanted in the ground or submerged in water. Measurement of the source to receiver time allows the speed or velocity of the acoustic waves to be calculated. Interpretation of the resultant velocities permits a geophysicist to make inferences about subsurface rock composition, rock competence (quality), and geological structure of the area surveyed. Estimated depths to the layers of different acoustic contrasts can also be calculated from the velocities and the geometry of the source and the receiver.

Seismic methods include the following:

- Seismic reflection, a method of reflecting acoustic energy from deep subsurface interfaces, principally used in oil exploration
- Seismic refraction, a method of refracting acoustic energy typically used to determine bedrock depths, physical properties, and rock quality at shallower depths than reflection surveys

Owing to the long wavelength with seismic signals required to penetrate the subsurface, definition (resolution) of small objects is poor. Seismic surveying is also labor and time intensive, and therefore, seismic methods have found very limited use in archaeological and forensic investigations.

Sonar surveys (i.e., side scan sonar), on the other hand, are well suited for marine investigations and are capable of producing high-resolution data of and within water bottoms. They have been used quite successfully in archaeological and forensic investigations.

3 Presurvey Planning and Postsurvey Reporting Requirements

Before a geophysical survey is to be performed, the forensic investigators should possess a working knowledge of the following factors.

3.1 Overall Statement of Objectives

Review of available site information to provide a descriptive assessment of the site:

- Target size and depth
- Proposed survey methods and survey locations
- Time constraints

Site-specific conditions are as follows:

- Surface: location, size, access, vegetation, climate
- Subsurface: soil types, groundwater, geology
- Sources of geophysical noise
- Site ownership, warrants
- Previous investigations

Integration of additional investigations includes the following:

- Aerial photography, historical and current
- Cadaver dogs
- Soils and vegetation
- Scavenging
- Previous excavations
- Historical weather records

3.2 Formulation of the Survey Plan: Questionnaire

Table 3.1 presents a questionnaire designed to provide information that can be utilized in the search for clandestine graves or evidence. Much of the basic information on the questionnaire will be useful for investigators to develop search strategies or a search footprint.

Table 3.1 NecroSearch Questionnaire

Save questionnaire as a new document on your computer before completing.

NECROSEARCH CLANDESTINE GRAVE QUESTIONNAIRE
(copyright NecroSearch International 1989, revised 2015)

NecroSearch is a team of civilian and sworn specialists whose aim is to assist law enforcement agencies in the detection of and recovery of evidence from clandestine graves. *All information supplied on this questionnaire is strictly confidential.* ***Please advise NecroSearch of any special needs or concerns you have in your investigation.***

NECROSEARCH HAS THE RESOURCES TO ASSIST YOU WITH ALL PHASES OF YOUR INVESTIGATION, INCLUDING TIMELINES, RESEARCHING BUILDING "AS-BUILT" PLANS, WEATHER REPORTS, LOCATION OF MAPS, UTILITIES, HISTORICAL AERIAL PHOTOGRAPHS, ETC.

PLEASE CONTACT US FOR HELP!

Requesting Agency:
Agency Case Number:
Contact Person:
Address:

Phone Number: Ext:
Alternate Phone Number: Ext:
FAX Number:
E-mail Address:

This Block for NecroSearch Use
N.S. Case No.________________
N.S. Contact________________
Date Presented
Actions Taken:

Note: We realize that you may not have all of the information requested on this questionnaire. For justifications for the questions, check notes at the end of this questionnaire.

You may wish to provide estimates or educated guesses, but please identify them as such.

1. **Information based on:** ___Witness statement ___Suspect(s) statement/confession ___Informant ___Anonymous source ___ Other (specify):

2. **Victim Information:**
   Name: Date of Birth: Sex:
   (Use separate sheet for each victim)
   Height: Weight:
   Any type of artificial medical devices?[a] Any other metal you suspect buried with body?
   Drug Use[b]:
   What clothing do you think the victim was wearing?

3. **Suspect #1 Information:**
   Name: Age: Sex:
   Height: Weight:
   Any physical disability or impairment:
   Is suspect capable of carrying victim? Yes No
   If so, how far in the suspected terrain (e.g., more or less than 50 yards)?

If there are additional suspects, please add them here with height, weight, etc.

(*Continued*)

Table 3.1 (Continued) NecroSearch Questionnaire

NECROSEARCH CLANDESTINE GRAVE QUESTIONNAIRE Page | **2**

4. How well did the suspect(s) know the area and possible hiding places?[c]

   Are there other areas and hiding places well known and frequented by these suspects?

   Have you eliminated these other areas from consideration?

5. What was suspected date and time the victim was killed (is this an estimate)?

6. Where was the suspected site the victim was killed (is this an estimate)?

   Where is the suspected site of burial or concealment (Latitude/Longitude/Street address)?

   What is the distance between the last known victim location and suspected site of burial?

7. What was the reported time interval between death and burial or concealment?
   What type of light was available (moon, sun, street-lights, etc.). **We can help with this.**

8. How did the victim reach the suspected site or area? Is this an educated guess?

Walked under own power?	
Carried?	
Dragged?	By how many individuals?
By Vehicle (inc. ATV)?	Make/Mdl/Color?
By Boat?	Make/Mdl/Color?
Other:	

9. Suspected access to area?

10. Are there any safety hazards our personnel should know about?

11. What is the terrain in the suspected area?

12. Was the body suspected to have been disposed of in an area to be covered over by construction[d]?
    If yes, explain:

    Please give us construction drawings (as built) of any building on the site if you suspect that the body may be hidden in and/or around them. We can help look this upif you wish!

13. Was the body suspected to have been disposed of in an area of heavy vegetation or trees[d]?
    If yes, explain; please describe:

14. Was the body suspected to have been disposed of in water[d]?

    If yes, explain; please include the water body type (stream, river, lake, reservoir, wetland, well, etc.) and name:

    If yes, was the body suspected to have been weighted?

    If so, what *specific* type of material was suspected to have been used?

    What other types of water structures/facilities (i.e., dams, canals, irrigation turnouts, pipelines, bridges, wells, etc.) are located on or around site?

(Continued)

Table 3.1 (Continued) NecroSearch Questionnaire

NECROSEARCH CLANDESTINE GRAVE QUESTIONNAIRE Page | 3

15. What were the ground conditions in this location at the time of the suspected concealment or burial (e.g., rocky/sandy)?

16. Have weather records been obtained?
    What weather records are available for review from the time frame of the suspected concealment or burial?
    We can help you obtain these records if you do not have them.

17. Do you suspect that a grave was dug?
    If not, how do you suspect the victim was concealed?

18. If you suspect a grave was dug, what tools were suspected to have been used?

    Have any of those tools been recovered?
    If so, have those tools been analyzed and are those reports available?
    How many individuals were suspected to have been involved in the digging?

19. Was earthmoving equipment suspected to have been used?
    (If you suspect the use of this equipment, it may be possible to eliminate some sites by having a person knowledgeable with that specific equipment review the suspected site)
    If so, what type of equipment?
    Has any equipment been recovered?
    If so, has this equipment been analyzed and are those reports available?

20. What type of covering do you suspect was used (soil, rock, brush, logs, lumber, concrete)?

21. Has anything happened since burial or concealment to alter the grave site (flooding,construction, fire, paving)?

22. Do you suspect the body was wrapped, if so how?

23. Has any physical evidence been recovered in the general area that is or may be associated with the victim(s)?

24. What scavengers (including birds) are in the area?
    What kinds of evidence (e.g., scat, hair, nests)?

(Continued)

Table 3.1 (Continued) NecroSearch Questionnaire

NECROSEARCH CLANDESTINE GRAVE QUESTIONNAIRE

25. What vegetation was believed or reported to have been present at the time of excavation?

26. Who owns the property to be investigated?

    Has consent been obtained to enter and work the scene?

    Has a search warrant been obtained?

    Is the owner/manager/caretaker available to give history and details?

    What is the history of the site?

    What information exists on location of known underground interferences (dumps, wells, septic and sewer systems, tanks, pipelines, animal graves)?

27. What maps of the area are available? **We can help you with this.**

    Topographic   Forest Service   Bureau of Land Management (BLM)

    National Park Service   Soil Conservation Service

    Real Estate Developers   Local Utilities

    Others

28. If outside the United States, what land/resource management agencies are available? **We can help you with this.**

29. What resource bases exist in your local area? **We can help you with this.**
    (Specialists at universities, colleges, utility companies, private firms, and equipment such as aircraft, cameras, geophysical equipment, and tracking/cadaver dogs?)

30. Are pre- and post burial aerial photographs available?

    How many years before burial or concealment?

    How many years after burial or concealment?

    What types? Color B&W Infrared Stereo Digital

    Source? US Government City & County Tax Assessor Highway Department Pipeline Company Railroad Public Service Private

    We may be able to help you obtain these photographs. However, photographs taken by your agency are extremely important!

31. Is an aircraft available to use for site photography?

    For site viewing?

32. Are site photos available? If not, can they be taken?

    Are photos available from or can they be taken:

    Early morning around sunrise?   Late afternoon around sunset?

(Continued)

Table 3.1 (Continued) NecroSearch Questionnaire

NECROSEARCH CLANDESTINE GRAVE QUESTIONNAIRE Page | **5**

Elevated position (e.g., tree, buildings, fire dept. ladder truck, etc.)? Ground level?

Photos of surrounding areas and access into site (roads, paths)?

33. Has any technical examination of the area been performed by others (foot searches, digging, geophysical surveys, dog searches, search and rescue, etc.)? If yes, describe briefly what was done and what results were obtained.

    Please provide the name(s) of the experts used and copies of their reports.

34. What reasons do you have to believe that the area in question is the area containing the grave? Is there any other information that you could offer that helps make this gravesite and grave unique and that would assist us in locating it? Our experience shows that some killers dispose of victims in unique ways. An understanding of disposal methods will help us to evaluate your case.

35. We have discovered that a timeline of events can be crucial in solving a case. If possible, please provide a synopsis and time line of the events. **We can help with this.**

Finally if you are successful in your search for a clandestine grave, we would appreciate receiving information on the particulars of the grave for our database. Information such as distance from the nearest town, nearest road, upslope or downslope, and type of covering, will help us all in evaluating and assisting in future cases.

PLEASE FEEL FREE TO CONTACT:

Diane L. France, Ph.D., D-ABFA DLF@HumanIDLab.com
President, NecroSearch International, Inc.
1713 Willox Court, STE A
Fort Collins, CO 80524
(970) 227-6514

Julia Bauer, Vice President jbauer@chillico.com

EXPENSES: NecroSearch members do not charge for their time. The **agency** requesting NecroSearch assistance will pay for all expenses incurred when NecroSearch members conduct a search, and will pay for travel and lodging expenses. Mileage within the United States will be reimbursed at US government mileage rates.

[a]We need to know of any metal buried with the body for remote sensing technology.
[b]Drug use is suspected to affect scavenging patterns.
[c]Research shows that bodies are hidden more often in locations owned or known by the perpetrator. Known areas are where the suspect uses for recreation such as hunting, fishing, and camping, as these may be areas within the suspects comfort zone for body disposal.
[d]These things affect ground-penetrating radar and other remote sensing technology.

*Courtesy of NecroSearch International

3.3 Presurvey Site Evaluation

Owing to the limitations of geophysical surveying, as much site information as possible should be gathered prior to conducting any surveys.

Other information, for example, potential sources of geophysical noise—such as fences, buildings, and power lines—should be identified from air photographs, maps, or visual reconnaissance. For example, the US Department of Agriculture, National Resource Conservation Service (USDA NRCS) has published ground-penetrating radar (GPR) suitability maps of most of the contiguous states. These maps provide an idea of the suitability and effectiveness of GPR in areas based on soil attributes. Soil attributes have been collated from State Soil Geographic and Soil Survey Geographic databases. These databases have also been used by USDA NRCS to produce state-specific soils maps (https://www.nrcs.usda.gov/wps/portal/nrcs/detail/soils/use/maps/?cid=nrcs142p2_053622).

A check with a local USDA NRCS soil scientist can provide very definitive data about a site of concern.

Geophysical measurements are generally made in a series of parallel lines or a cross-grid pattern across the survey area. Grid lines can be laid out with rope or tape, and grid stations along each line can be marked with flagging on plastic pin stakes. The size and depth of the objects sought determine the spacing of measurement stations or survey lines. In forensic surveying, line and station spacing may be on the order of 0.5–1 m or less.

Performing a test survey over a known or a constructed feature is highly recommended. For example, in searching for a clandestine grave, a similar grave could be constructed if the details are known. This provides an opportunity to observe the type of response the instrument chosen for the survey will give. This type of calibration survey will also provide information on the optimum line spacing and station interval to use in a grid to be able to detect a specific target.

Concurrent with the geophysical survey, a site features map should be developed to document the location of surface scrap metal, power lines, buried utilities, roads, topographic and cultural features, and soil and vegetation changes observed at the site. This information will be used during the data interpretation stage to provide a better understanding of the significance of geophysical anomalies detected. A site features map also offers an excellent visual presentation for a jury.

3.4 Equipment

Electronic geophysical equipment, often subject to field and travel-related abuse, is susceptible to different problems, all of which may affect the quality of recorded data. All equipment, when supplied by the manufacturer, is calibrated to a nationally or professionally accepted standard. Equipment calibration often diverges from these standards owing to age and use; thus it is a standard, professional practice for geophysicists to perform field calibrations prior to the start of a survey. It is recommended that the make, model, and serial number of each piece of equipment used in a survey be recorded along with the data and place of the last manufacturer's calibration. It may be possible to perform a survey when an instrument is known to be out of calibration, but the results will not stand up to legal scrutiny.

3.5 Personnel

The complexity of an investigation will dictate the level of personnel education and experience necessary to perform a geophysical survey. Where a single geophysical method is proposed for an investigation, a qualified geophysicist specializing in that method should be used to perform that survey. Where multiple methods are used, an organization providing comprehensive geophysical expertise should be used to perform the surveys. Instrument operators who are not trained geophysicists are typically less experienced in data reduction and interpretation routines than trained geophysicists. However, nongeophysicists may be trained and qualified to operate specific instruments under the supervision of a qualified geophysicist.

There are 29 states and 1 territory with geologic registration or licensure requirements for geologists, geophysicists, or geoscientists. Some of these states require that geophysical surveys be performed by, or under the supervision of, a licensed or registered professional. The results of a geophysical survey performed by an unregistered individual in a state that requires registration may not be able to be used in a court of law.

3.6 Reporting

A written report of findings should include as part of the case file copies of the geophysical profiles and/or contour maps depicting the results of the survey. The following items should be discussed in the report: description of the site, description of the survey procedures and equipment utilized, survey grid parameters, an interpretation of the results and their significance, and any conclusions and recommendations. The report should contain all data sheets, computer printouts, plots, and the site features map.

4 Magnetic Surveying

4.1 Introduction

Magnetic surveys are primarily used to detect ferrous metals, iron, and steel. The earth's magnetic field (geomagnetic field) has the shape of a bar magnet and exhibits an intensity that ranges from 70,000 gammas (γ) [nanoteslas (nT)] at the polar regions to 25,000 γ at the equator. The field is shifted some 11° from the earth's rotation, which causes a deviation from true north, resulting in a point referred to as magnetic north. Where local variations in magnetic materials occur, anomalies within the total magnetic field of 1–5000 γ may result.

Detection of anomalies is a function of the differences in contrasts owing to the material's magnetic susceptibility; remnant magnetization; size, shape, and orientation of the magnetic feature; and the distance between the feature and point of measurement. Magnetic susceptibility is the ease with which a material can be magnetized by the earth's field or an induced field (such as a magnet). Remnant magnetization is the natural or permanent magnetization displayed by a material in the absence of a magnetic field.

The significance of magnetization to the forensic investigator is that most weapons have a high magnetic susceptibility, and thus they may be capable of being detected by magnetic surveys. Soils display remnant magnetization in the first few feet below the ground surface. Should soils be altered by excavation, such as for a grave, their remnant magnetization will also be altered, resulting in a magnetically anomalous condition that often can be detected by careful magnetic surveying.

Magnetic measurements are made by utilizing an instrument called a magnetometer. Table 4.1 shows typical magnetic responses for various materials based on the distance of the material from a magnetometer.

Table 4.1 Magnetic Responses

Typical Maximum Anomaly		
Object	*Near Distance*	*Far Distance*
Automobile (1 ton)	30 ft = 30 γ	100 ft = 40 γ
Light aircraft	20 ft = 10–30 γ	50 ft = 0.5–1 γ
Pipeline (12 in. diameter)	25 ft = 50–200 γ	50 ft = 10–50 γ
Metal fence line	10 ft = 15 γ	25 ft = <5 γ
Rifle	5 ft = 10–50 γ	10 ft = <10 γ
Revolver (0.38 special or 0.45 automatic)	5 ft = 10–20 γ	10 ft = <5 γ
Metal file (10 in.)	5 ft = 50–100 γ	10 ft = <10 γ
Screwdriver (5 in.)	5 ft = 5–10 γ	10 ft = <5 γ

4.2 Limitations

The type of target commonly sought by the forensic investigator, weapons, and excavations results in relatively small magnetic contrasts in the subsurface. Magnetic "noise" owing to cultural features, geologic conditions, and solar activity can mask these small contrasts:

- Cultural features include electrical power lines, pipes and pipelines, metal structures both underground and above ground, fences, metal signs, automobiles, and metallic surface trash. High noise levels can make interpretation difficult and, in some instances, can make data collection and interpretation impossible.
- Geologic conditions: Magnetic surveys are affected by rocks and soils having high magnetic susceptibilities or remnant magnetism. Igneous and metamorphic rocks contain relatively high volumes of the mineral magnetite. Soils derived from them may have concentrated amounts. Highly organic soils produce maghemite, the magnetic form of the mineral hematite.
- Solar activity: Solar winds and solar magnetic storms (resulting in sunspots) can greatly affect results obtained from magnetic surveying. Daily solar activity, referred to as diurnal change, can produce spurious anomalies on the order of 100 γ in the course of a day's surveying. The National Weather Service, the Solar Forecast Center (Boulder, CO), and the National Geophysical Data Center (Golden, CO) can provide predictions of solar magnetic activity.

The use of a base station magnetometer and/or the magnetic gradient surveying method diminishes or eliminates the effects of normal diurnal changes.

4.3 Instrumentation

The type of magnetometer best suited for a particular site investigation depends upon the characteristics of that site and should be chosen by a person familiar with the different instruments available. The output of a magnetometer is a numerical value (γ) of the intensity of the earth's magnetic field at a single location, such as a survey station.

Different instruments have different levels of sensitivity. In some cases, high sensitivity may be desired to detect deeply buried or very small objects, and in other instances, a low sensitivity instrument may be desired to reduce the effects of "noise" from nearby objects such as fences or cars. Proton magnetometers, for example, although very useful in some situations, will cease to function accurately in an area with high magnetic gradients such as a junkyard or near a steel bridge.

The time of each magnetic reading must be recorded to calculate the diurnal drift correction. Some magnetometers include microprocessors for recording and storing measurement values of readings, the line(s) and station locations, and reading times. Magnetometer memory storage capabilities depend on the type of instrument used in surveying. The less sophisticated (less expensive) models will record 1,200 data points while more sophisticated instruments can record over 100,000 data points.

Because the objective of magnetic surveying is often to complete a magnetic map of the survey area, each magnetic reading must be plotted at its precise location. Depending on the site and survey requirements, an instrument operator can acquire between 1,800 measurements a day with a conventional instrument and 15,000 measurements a day with a continuously sampling instrument.

4.3.1 Total Field Magnetometers

The proton precession magnetometer is commonly used because of its availability, ease of use, and satisfactory measurement precision. It measures the absolute total intensity of the magnetic field to a 0.1 γ precision. Figure 4.1 shows a typical proton magnetometer.

Figure 4.1 Geometrics proton magnetometer.

Overhauser magnetometers are capable of running at higher speeds than proton magnetometers, and therefore can be used to produce almost continuous measurements. Overhauser magnetometers can be used in higher gradient magnetic fields than proton magnetometers. The instrumentation package appears similar to the proton magnetometer as shown in Figure 4.1. Overhauser magnetometers are capable of achieving an absolute accuracy of 0.01 γ.

Fluxgate magnetometers are capable of continuously measuring the relative changes in the earth's field. The instrument requires orientation with the earth's field; however, this allows the vertical and horizontal magnetic components of the earth's field to be measured. Sensitivities of 0.1 γ are obtainable with fluxgate magnetometers. The primary advantage of the fluxgate magnetometer is the capability of taking continuous readings. Figure 4.2 shows a typical fluxgate magnetometer.

Optical absorption magnetometers are also referred to as cesium vapor magnetometers. These instruments can obtain a sensitivity of 0.1 γ, and are capable of detecting a 55 gal metal drum at a depth of 30 ft, or 1 lb of iron at a depth of 10 ft. Cesium vapor magnetometers are capable of taking 10 measurements per second; thus, they can

Figure 4.2 Fluxgate magnetometer.

be used to survey areas at a much quicker rate than surveying with a proton magnetometer. Figure 4.3 shows a typical cesium vapor magnetometer. Cesium vapor magnetometers are typically used for the detection of unexploded ordnance.

4.3.2 Gradient Magnetometers

The measurement of the magnetic field utilizing two magnetic sensors mounted in the vertical or horizontal mode is known as gradiometer surveying. Figure 4.4 shows a proton magnetometer set up as a gradiometer.

A gradiometer is used to measure changes per unit distance of the earth's magnetic field, with units of measurement in gammas per foot or gammas per meter. The advantage of gradiometer surveying is that the use of two sensor heads minimizes the effects of strong magnetic gradients from cultural features and surface debris, diminishes diurnal effects, and removes the effects of a regional magnetic gradient.

Figure 4.3 Cesium vapor magnetometer.

Figure 4.4 Gradiometer, two sensor heads.

Because the sensors are separated by a small fixed distance, typically 2–3 ft, the difference between the measurements will be very small. Subtraction of the magnetic measurements from the two sensors may help define near-surface anomalies with increased resolution. The magnetic gradient surveying method is preferred for forensic investigations.

4.4 Field Procedures

Prior to magnetic data collection, the endpoints of each survey line should be staked and marked. The location of these points should be accurately transferred to a base map—this is vitally important because when the field data are plotted on the base map any anomalies will have to be located in the field based on their accurate location on the base map. Nonmetallic stakes should be used for all field operations. When possible, survey lines should be oriented in a north–south direction. An example of a survey grid is shown in Figure 4.5.

Prior to starting the survey, the instrument batteries should be checked for sufficient charge. The magnetometer should also be tested/calibrated according to the manufacturer's operating procedures. Calibration of the magnetometers consists of "tuning" or adjusting the sensitivity with reference to the magnetic field at the site. This magnetic field across the United States is approximately 50,000–60,000 γ. The operational manuals that come with magnetometers have maps of worldwide magnetic field strengths that can be used for tuning. Tuning is conducted in an area free of magnetic disturbances if possible.

To initiate the survey, the instrument operator proceeds along each survey line taking a measurement (reading) at each station. Generally, at each survey station, the button on the instrument is depressed, which

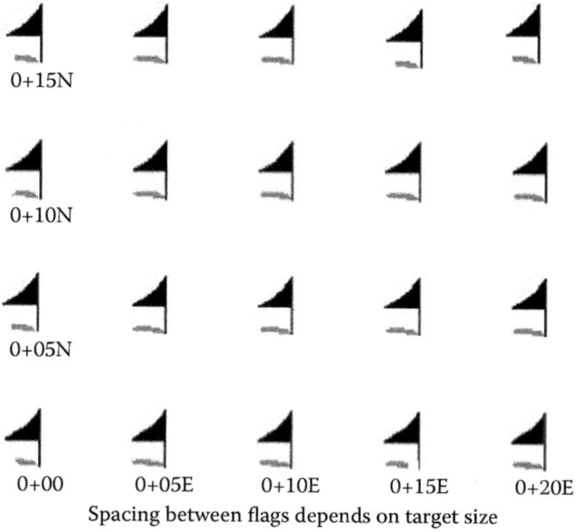

Figure 4.5 Geophysical survey grid.

results in a data measurement. The measurement obtained at each station will be recorded in a field log book and/or in the instrument memory. The line and station number should also be recorded.

The operator then advances to the next station and obtains a reading. This process is continued until measurements have been obtained at each station within the survey grid. With a fluxgate magnetometer, or "walking"-type proton magnetometers, the operator adjusts his pace such that the instrument takes automatic readings as the operator walks along each line. Continuous sampling with a "walking magnetometer" is the preferable method of collecting data in forensic investigations.

Normally, when using noncontinuous sampling instruments, three consecutive magnetic readings are taken at each reading station. If the readings agree to within 2–3 γ, they are considered to be valid. If there is a large variation in the readings, they are not considered valid, but most likely represent the effects of noise. If the effects of noise cannot be removed in the field (a small displacement of the station location, for instance), the readings should be noted as invalid.

Owing to natural diurnal magnetic variations, corrections are made to the field data. The optimum solution for obtaining measurements of diurnal changes is to use two magnetometers, one of which remains in a fixed location (the base) and continually records data during the course of the field survey. This data can then be used to make the necessary diurnal corrections. When the use of a base station magnetometer is not possible, to determine the time-varying diurnal effects, the operator should make repeated readings at a fixed location (base station) at different times during the course of the survey. When performing gradiometer surveys, it is not necessary to make diurnal corrections.

4.5 Special Considerations

It is vitally important that the instrument operator be as free as possible from magnetic material, as the presence of such material can adversely affect the magnetic data. The operator should be free of objects such as keys, belt buckles, metal zippers, and steel-toed boots. Measurements must not be taken with the sensor near ferromagnetic objects. The orientation of the magnetometer sensor, its height above the ground, and its maintenance in a vertical position must be carefully controlled while surveying.

4.6 Documentation

Good field notes are necessary to indicate variations in survey parameters such as line location deviations and station skips. A field book is generally kept with the operator to record these observations. The field book is nearly always referred to during data reduction and analysis to help explain some inevitable data discrepancies. It is imperative that excellent notes be kept in legible form, because the notebook will become part of the chain of evidence.

Data stored within a memory magnetometer can be downloaded to a computer such that standardized data sheets are not necessary. Information obtained during a magnetic survey that is not stored within a magnetometer should be recorded in a field log book. In the event that magnetics (MAG) data are not recorded, a field log book should be used to record survey information. This information should include agency case number, site location [ground positioning system (GPS) or address], date, weather, instrument make, serial number, and last calibration date. Survey information would include station number (in not recording continuously), line number, station number, magnetic reading, comments, a sketch map, and the operator's signature.

4.7 Data Reduction

Before magnetic data may be interpreted, it must first be corrected for regional magnetic gradient and diurnal drift. Because most sites involved in forensic investigations are quite small, the regional gradient will have a negligible effect on the survey data. Diurnal drift—which is not a function of the size of a site but rather of the amount of data collected over a time period—is corrected by calculating the base station field strength changes over time and subtracting the appropriate time-dependent value from each observation. The correction for each reading may be determined by drawing a smooth curve through the base station readings. These readings in gammas (y-axis) are plotted versus the time of each reading (x-axis). The value of the correction (γ) per time interval is determined from the smooth curve. This value is then multiplied by the difference in time between each successive field reading, and the result is subtracted from the field reading. This process is completed for each field reading. Newer magnetometers contain software to calculate corrections and produce corrected magnetic data for each measurement station.

Once the diurnal corrections have been made, the corrected magnetic survey data are plotted in the form of a profile or as a contour map.

4.8 Data Interpretation

Typically, data reduction and interpretation will require at least an equivalent amount of time to perform as was used to acquire the field data. This is the 1:1 rule: 1 day of surveying results in 1 day of analysis and interpretation; 1 hour of surveying results in 1 hour of analysis and interpretation.

Preliminary interpretation of the geophysical data is normally performed while in the field and immediately after completion of a survey grid. This aids in planning remaining work or modifying the survey program. Magnetic anomalies are interpreted with respect to their amplitude, lateral extent, and shape. Depth estimates of the magnetic source(s) can be made based on these anomaly characteristics. Owing to the inclination of the earth's magnetic field, anomalies often will not appear directly above their source(s). In forensic surveying, very often the preliminary interpretation is sufficient to solve the problem of detection and delineation of a target.

Should a detailed interpretation be necessary, the process involves selecting a number of possible models that may produce the type of magnetic anomaly measured in the field. There are a number of computer programs available to interpret the size, shape, and depth of a magnetic source(s) based on the magnetic anomalies in the field data.

4.9 Data Presentation

The magnetic field data are normally presented either in profile form or as magnetic intensity contour maps. When in profile form, the magnetic intensity, in gammas, is plotted on the y-axis and the line stations are plotted on the x-axis (Figure 4.6). The data can also be presented in the

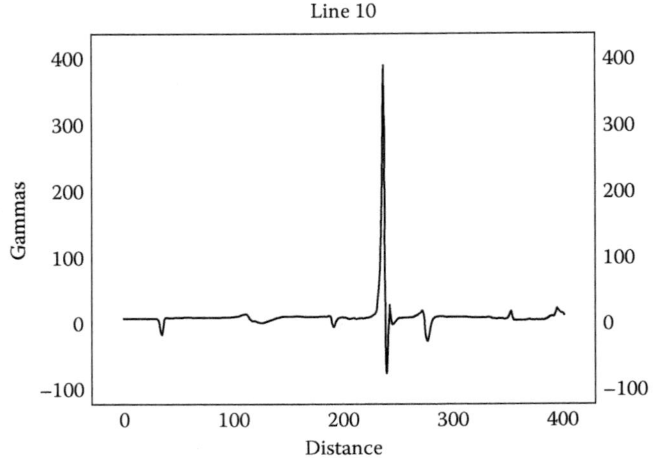

Figure 4.6 Magnetic profile.

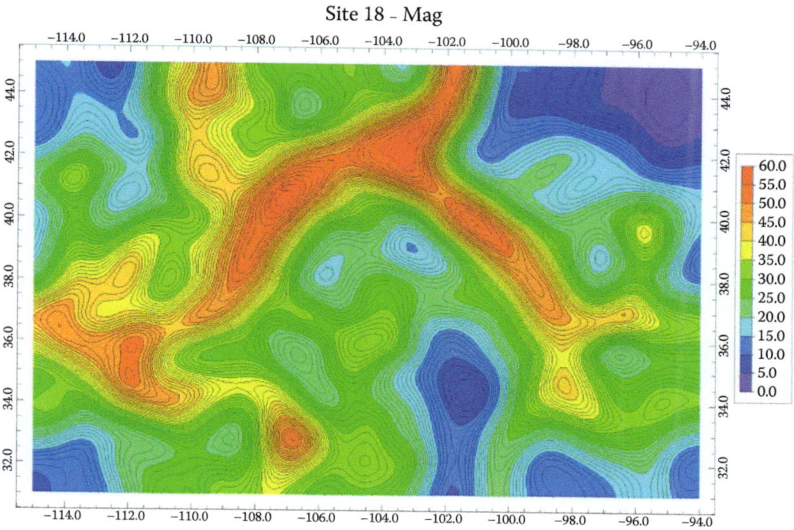

Figure 4.7 Magnetic map.

form of map (Figure 4.7). Magnetic maps are quite useful in illustrating magnetic features in an area of interest, although subtle features that can often be noted on profile plots may not be evident on contour maps.

4.10 Summary

Magnetic surveying can be used to:

- Detect ferrous metal, that is, any metal that will rust
- Detect disturbed soils
- Detect material whose magnetic properties have been altered by heat

Magnetic surveys can be affected by:

- Interference from metallic objects, for example, fences, power lines
- Solar activity
- Geologic conditions

5 Electromagnetic Surveying

5.1 Introduction

Magnetic surveying and electromagnetic (EM) surveying are considered to be complementary techniques, in that magnetic surveying is used for the detection of ferrous metals, while electromagnetics is used for the detection of ferrous and/or nonferrous metals.

The EM method provides a rapid means of measuring the relative changes in conductivity between buried metallic objects, subsurface soil and rock, by the induction of an EM current into the subsurface. A small alternating current passing through a transmitter coil produces a primary, time-varying magnetic field within the ground. Through inductive coupling, the primary magnetic field produces small eddy currents in the subsurface which, in turn, create their own secondary magnetic field (Figure 5.1). The receiver coil senses both the primary and secondary fields.

Utilizing a nonsurface contacting transmitter–receiver arrangement, the interaction of the generated circular eddy current loops or EM field with earthen materials is directly proportional to the terrain conductivity within the influence area of the instrument. Conductivity, measured in units of siemens or millimhos per meter (mmho/m) of material, is the reciprocal of resistivity. Therefore, the information provided by a conductivity survey should, in theory, produce similar results to that of a resistivity survey. The main advantage of terrain conductivity surveying over that of resistivity surveying is that the instrumentation does not have to be in contact with the ground surface because the EM signal is inductively coupled to the subsurface.

Changes in the magnitude and the phase of the individual currents are generally related to the terrain conductivity. Terrain conductivity is a function of the soil or rock type and composition, the porosity and permeability of the subsurface units, the conductivity of the fluids

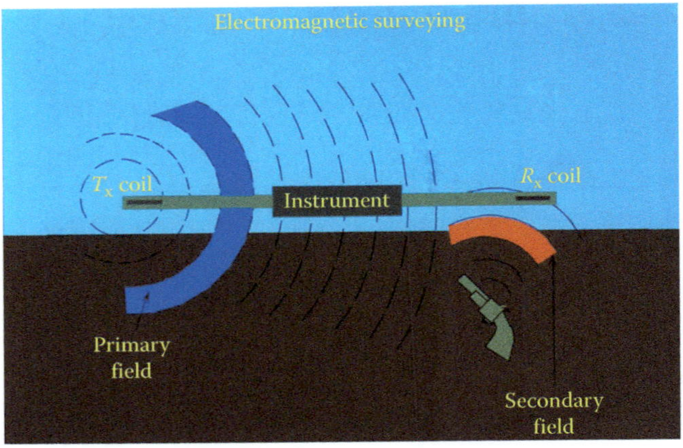

Figure 5.1 Electromagnetic field.

Table 5.1 Conductivity Variances as a Function of Materials

Process/Object	*Conductivity*
Backfilled excavation	Increase
Mineral-rich soil	Increase
Salt water	Increase
Clay increase	Increase
Wet soil	Increase
Metallic objects[a]	Increase/decrease
Fertilized soil	Increase
Silicification	Decrease
Mineral-leached soil	Decrease
Soil compaction	Decrease

[a] The shape of a metallic object and its orientation to the EM field will determine if a conductivity increase or decrease will exist. For example, when performing EM surveying parallel to metal pipes, a high conductivity will be evident, while when surveying perpendicular to the same pipes, a low conductivity will be evident.

filling the pore spaces, and the presence of buried conductive (metallic) objects. EM methods are very sensitive to subsurface features, such as lateral soil changes owing to compaction of backfill, buried metallic objects, and ancient habitation sites.

Table 5.1 can be used as a guide for determining the effects on conductivity for different objects or geologic conditions.

5.2 Limitations

EM measurements taken over an area are an average of all ground conductivities within the depth range of the instrument used. The depth range is a function of the transmitter to receiver coil separation and frequency of the transmitted signal.

EM interference from surface cultural features will be averaged into the measurement. These features, such as buildings, fences, and buried objects and utilities, must be accounted for by maintaining a set distance between the feature and the instrument. A trained operator will be aware of different signal responses from subsurface cultural features such as underground utility lines. Field tests will indicate at which distances features will interfere with EM measurements. Areas exhibiting high conductivities, such as clayey soils, may not permit much penetration of the signal owing to dissipation of the EM field.

5.3 Instrumentation

Currently available instrumentation consists of two basic types. One type, which is almost always used for forensic investigations, consists of a transmitter coil and a receiving coil mounted together, at a fixed distance apart. The other type of EM instrumentation, primarily used for mineral and groundwater explorations, consists of separate transmitting and receiving coils that can be independently moved to achieve greater separation than fixed coil instruments. The extent or size of the induced EM field is determined by the coil separations.

Most conductivity equipment incorporates a voltage meter calibrated to read conductivity as a function of an output voltage. The output voltage is, in some cases, linearly proportional to the conductive coupling effects of the subsurface material within the zone of transmitter–receiver influence. A semi-spherical area around the transmitter–receiver coils is investigated, the volume of which is related to the coil separation distance and the orientation of the coils.

Forensic-related conductivity anomalies generally are near the surface, are relatively small, occupy a small volume, and are probably not very conductive, if they are of a nonmetallic nature. Therefore, equipment must be capable of providing a highly sensitive measurement of a relatively small volume area. To accomplish this, coil spacing must be small, perhaps 1–4 m in length. A 1-m coil spacing instrument, such as a Geonics EM38 (Figure 5.2), below the ground surface with the instrument resting on the ground surface, is capable of penetrating approximately 1.5 m (5 ft) in depth. A larger coil spacing, 3.7 m, is available in the

Figure 5.2 Ground conductivity meter, Geonics EM38 with extender arm. (Courtesy of Geonics Limited.)

Figure 5.3 Ground conductivity meter, Geonics EM31. (Courtesy of Geonics Limited.)

Geonics EM31 model, as shown in Figure 5.3. This instrument averages conductivity measurements over a much greater soil volume, to a depth of approximately 6 m (18 ft).

Utilizing a 1-m coil spacing instrument across a site on a 1-m station interval along each line will result in a significant number of measurements. In some investigative cases, e.g. for small discrete targets such as hand-held weapons, smaller station spacing may be required, further increasing the number of measurements. To facilitate such data recording, it is recommended that a digital recording device, such as a field computer (data logger) be employed. The acquired data can be downloaded to a computer upon survey completion.

5.4 Field Procedures

Prior to EM data collection, the endpoints of each survey line should be staked and marked. The location of these points should be accurately transferred to a base map. Nonmetallic stakes should be used to mark station location in the field. The orientation of the survey lines is not important, as it is in magnetic surveying. For reference, an example of a survey grid is shown in Chapter 4 (Figure 4.5). Simultaneous collection of GPS data during surveying may negate the need grid surveying.

Instrument calibration is normally set at the manufacturer's facility and should be checked periodically by the manufacturer. In the field, the instrument operator will normally establish a baseline, outside the grid to be surveyed, to tune the instrument based on actual site materials and to ensure proper instrument operation. Measurements along this line should be repeated periodically during the course of the survey. Measurements along this line will give the operator information on the amplitude and spatial changes that will effect EM measurements.

Initial checkout procedures for the instrument involve checking battery condition, zeroing the instrument at its least sensitive scale (typically in units of millimhos per meter), and checking the instrument sensitivity to ensure proper scale readings during the survey. Both the EM38 and the EM31 are capable of reading two components of the EM field, the in-phase and the quadrature phase components. The in-phase component is primarily sensitive to metals, while the quadrature phase is primarily indicative of changes in ground conductivities.

Data acquisition is a straightforward process, similar to magnetic surveys. Measurements can be made along a line on a station-to-station basis or continuously. In fixed coil (EM31 or EM38) surveying, the instrument is kept at a constant height above the ground, and parallel to the ground surface. This will limit interference owing to measuring smaller or greater volumes of the subsurface if the instrument is varied in height. The instrument may be maintained in a vertical coils orientation position. Vertical coil orientations offer, in general, lesser depth penetration than horizontally oriented coils. As a general rule, when using horizontally oriented coils, the depth of penetration is approximately 1–1/2 times the coil separation, and with vertically oriented coils, it is 3/4 of the coil separation.

At each station, the instrument should be consistently oriented in the same direction. Taking two readings at each station, perpendicular to each other, will identify lateral changes, if they exist. This is particularly important because a clandestine grave may exhibit a significant measurement variation if the orientation of the instrument is parallel to or perpendicular to the feature. It is usually more time-efficient to acquire data with the coils located with one orientation and then rerun the survey grid with the coils oriented in a direction perpendicular to the initial orientation.

Large negative or positive fluctuations in measurements over a small area may be indicative of highly conductive subsurface materials, possibly related to metallic objects. Some instrument manufacturers recommend determining the effects of nearby cultural features by moving away from a feature until its effects on the measurement are negligible. Besides cultural influences, the operator should be aware of electrical storm activity (spherics), as it may cause meter readings to fluctuate beyond acceptable noise levels. When spherics are a problem, the use of a horizontal (vertical coils) coil orientation will minimize their effects.

5.5 Special Considerations

Because fixed coil instruments are very sensitive, it is recommended that instrument operators remove all metal objects from their person prior to performing EM surveys. The operator should be free of objects such as keys, belt buckles, and steel-toed boots. Because of possible EM field effects during transmission, the operator should not carry a cell phone or as a minimum place the cell phone in the Airplane Mode during survey operations.

5.6 Documentation

Similarly to magnetic surveys, all measurements, line numbers, and station locations can be digitally recorded to a data logger. For small-scale surveys, a written record of line, station, and measurements will suffice. No matter which method of documentation is selected, survey comments should be explained with sufficient detail to aid in the interpretative process. These comments may include location of cultural interference, weather, and observed geologic features.

Information obtained during an EM survey can be presented using standardized data sheets. In the event that a data logger is not used to record EM survey information, a field log book should be used to record survey information. This information should include agency case number, site location [ground positioning system (GPS) or address], date, weather, instrument make, serial number, and last calibration date. Survey information would include line number, station number (in not recording continuously), operation mode (IP or Q), vertical and horizontal coil orientation EM station values, comments, a sketch map, and the operator's signature.

5.7 Data Reduction

All conductivity values—in millimhos per meter (or millisiemens per meter, mS/m) and in-phase values, in parts per thousand (ppt)—are subsequently plotted on a map and/or computer processed so that their variation over the site can be analyzed.

In most cases, very little data reduction is necessary when the primary purpose of the survey is to observe lateral and/or spatial variations rather than absolute conductivity values. Resistivity is the reciprocal of conductivity, in units of ohm-meters. Usually the resistivity of materials as measured by EM methods is slightly lower than the resistivity of the same materials determined using direct current equipment. This is because the measured conductivity of a material increases as the frequency of the measuring signal increases.

Generally, a plotted profile or a contour map depicting apparent conductivity and in-phase values versus station location is the final data reduction product. For forensic investigations, it is recommended that these plots be produced in the field, manually or by a computer.

5.8 Data Interpretation

The interpretation of spatial variations in conductivity values is normally a qualitative one. Because the objective of an EM survey is to detect changes from a relatively constant background, both positive (increasing) and negative (decreasing) conductivity can be of significance to a forensic investigation. Conductivity survey results across sites are often variable. Changes in subsurface moisture content and underground utilities are often the cause. Buried conductive metals can

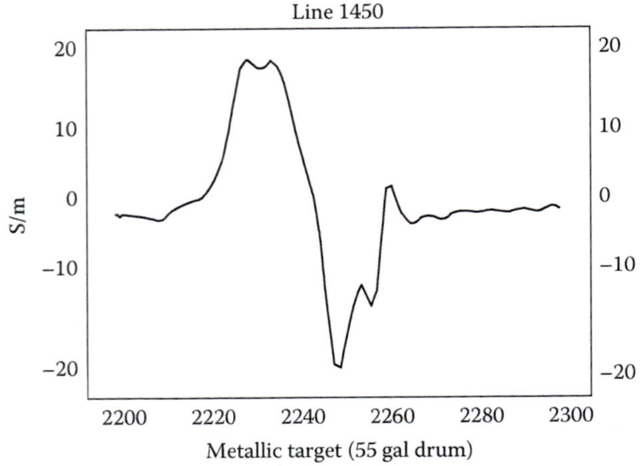

Figure 5.4 EM profile. (Author's collection.)

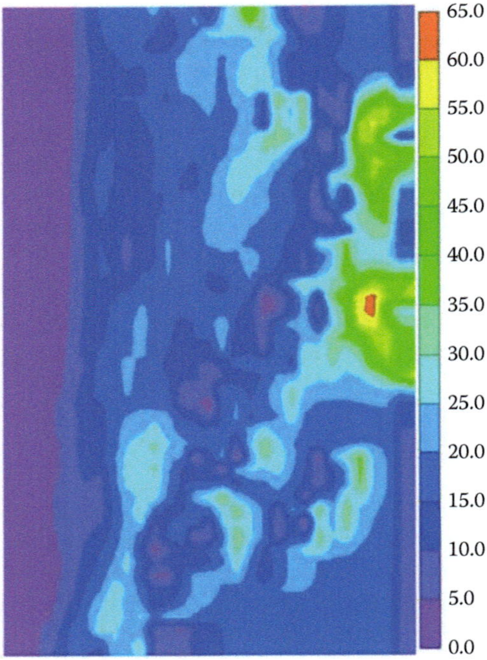

Figure 5.5 EM iso-conductivity map. (Courtesy of Rockware.)

produce positive or negative anomalies, depending on the shape of the source and its orientation with respect to the EM system coils.

5.9 Data Presentation

The EM field data are normally presented either in profile form or as iso-conductivity contour maps. When in profile form, the conductivity value in mS/m is plotted on the *y*-axis, and the line stations are plotted on the *x*-axis (Figure 5.4). For in-phase data, the ppt value is plotted on the *y*-axis and the line stations on the *x*-axis. Iso-conductivity maps (Figure 5.5) are quite useful in illustrating features in an area of interest. Computer processing can be used to enhance the iso-map.

5.10 Summary

EM surveying can be used to:

- Detect disturbed soils based on conductivity (moisture or chemical) changes
- Detect ferrous and nonferrous metal

Magnetic surveys can be affected by:

- Interference from cultural features, for example, fences, powerlines, and underground utilities.
- Geologic conditions

6 Ground-Penetrating Radar Surveying

6.1 Introduction

Ground-penetrating radar, often referred to as GPR, is an active electromagnetic subsurface investigative method. Its operation is based on the introduction of a relatively low-frequency electromagnetic signal, generally in the 10–1000 megahertz (MHz) range, into the ground via a transmitting antenna in contact with the ground. As this signal penetrates into the earth, it will be reflected, refracted, and/or diffracted as it encounters materials of contrasting electrical properties (Figure 6.1).

Where the subsurface consists of highly conductive materials (clay, saltwater, metallics), the radar signal will be dissipated, limiting the penetration.

The time between the transmitted and received reflected signal is a function of the velocity of the signal in the subsurface. The time it takes from signal transmission to reflection and then reception is referred to as two-way travel time, which is used to calculate the signal velocities as they penetrate into the earth. Once the velocity is calculated, the depth of different targets in the subsurface can be determined.

Penetration of GPR signals within the subsurface is dependent on the electrical properties of the materials and the frequency of the transmitting antenna used. GPR antennae frequencies of 10–1000 MHz are commercially available and will suffice for forensic investigations. Greater depth penetration is obtained from lower frequency antennae; however, higher frequency antennae offer more resolution (with less penetration). For example, in dry, sandy soils with little clay, an 80 MHz antenna may be capable of penetrating to a depth of 10–15 m, whereas a 1000 MHz antenna will most likely penetrate only 1 m or less.

The magnitude (amplitude), phase (negative or positive), and frequency of the received signal provide information of the nature of the subsurface materials. In many instances, a strong reflected signal

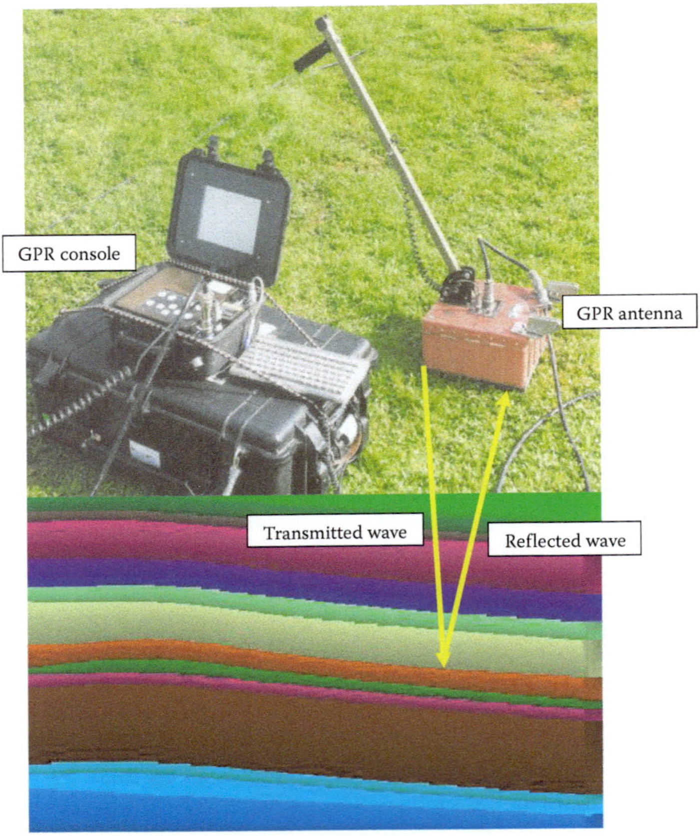

Figure 6.1 GPR profiling schematic. (Courtesy of Rockware.)

may indicate a buried metallic object. Undisturbed, layered soil will generally be observed on radar recordings as horizontal layers, while disturbed soils and subsurface materials will be indicated by erratic patterns.

6.2 Limitations

Water saturation and chemical activity generally control the electrical properties of the materials. High-conductivity materials will attenuate radar signals causing a decrease in depth penetration. The presence of small amounts of highly conductive mineralogic clay materials is the greatest limiting factor affecting the radar method. GPR surveys can be conducted over freshwater, but not over saltwater (due to its high conductivity).

If mineralogic clay or water saturation is suspected, GPR surveying may not provide useful data. Electromagnetic data may provide information on the conductivity of the materials to be surveyed with GPR. A general rule of thumb is that if the resistivity of the material(s) is $<30\,\Omega$m, GPR surveying will not provide useful data. The optimum location for a GPR survey would be an area of dry soil, flat, unvegetated surface with no obstructions. The author used GPR surveying at an optimum location in Metaponto, Italy, to delineate the grave of a soldier who was killed in 216 BC. Rarely do clandestine graves exist in optimum locations!

> Perhaps the most severe limitation of using GPR in forensics surveys is that GPR will not detect human remains. GPR manufacturers and geophysicists have battled misconceptions based on how GPR is portrayed in films and TV. Many law enforcement personnel seem to be under the impression that GPR is a silver bullet and that they will see a body or skeleton in the results of a GPR survey!

6.3 Instrumentation

There are two basic radar systems commercially available: a reflection profiling system and a pulsed profiling system. Both systems contain the same basic components: antennae, a waveform control module, and a video display unit. The reflection profiling system emits a continuous signal, while the pulsed system emits a rapid but noncontinuous signal. Data output from both systems is similar.

The GPR antennae vary in size, the lower frequency antennae being much larger than the higher frequency ones. The lower frequency antennae may be wheel mounted so that they can be towed by an operator. GPR antennae are not typically on wheels, but can be hand pulled or cart mounted and pushed. Some antennae have a trigger switch built into the pulling handle. When the antenna operator passes a grid flag, the switch is pushed, sending an electronic pulse to the control module, establishing a fiducial mark on the video displayed GPR data. Some systems allow the incorporation of a distance wheel and GPS which, when calibrated, establishes a fiducial mark on the video displayed GPR data. In this manner, features on the GPR record can be accurately related to the survey grid. Figures 6.2 and 6.3 show some GPR reflection profiling systems.

The waveform control module is used to optimize the radar signal that is being transmitted and enhance that received radar signal. Control modules typically contain gain and filter settings. Gain controls

Figure 6.2 Radar profiling systems, GSSI cart-mounted system. (Courtesy of Geophysical Survey Systems, Inc.)

are used to modify the amplitudes of transmitted and/or received signals, while filter controls are used to enhance the received signal by filtering instrument and cultural noise.

Recorded radar data can be played back at different speeds, stretched or compressed to enhance anomalies, and processed by computer software programs.

6.4 Field Procedures

For some field GPR surveying purposes, a two-person crew, an instrument operator, and an antenna operator are required to operate the equipment. However, with newer cart-mounted or backpack systems, only an instrument operator is used, once the survey grid is laid out.

Instrument calibration procedures involve optimizing the time–depth range of the instrument as determined by anticipated target depth and

Figure 6.3 Radar profiling systems, GSSI pulled system with distance wheel. (Courtesy of Rockware.)

by adjusting the signal gains. Radar signal loss or dissipation increases with depth. Random noise also increases with depth, often obscuring reflectors of interest. Adjusting signal gains and filter settings provides a means of enhancing or increasing reflected signal strengths from deeper occurring interfaces or anomalies.

For actual data acquisition, a survey grid of parallel line intersecting sets is established across a site with line separations determined by the dimensions of the target being sought. Along each line, location marks will be situated at station intervals, generally on the order of 1–3 m.

When using a survey wheel, distance marks along each line are automatically placed on the video-recorded data. Most systems that law enforcement will encounter will likely be using a survey wheel and GPS. Should a survey wheel not be available, the antenna operator actuates the location marker switch as the center of the antenna passes over each station location.

The towing speed of the antenna will vary depending on site conditions and topography, but generally should be on the order of 0.5–1.0 mph. Variable speed will cause stretching or compressing of the data, but this can be compensated for in the data processing and interpretation phase.

6.5 Special Considerations

Some older, low-frequency antennae (below 300 MHz) are unshielded and radiate energy in all directions. With these antennae, above-ground objects—such as large tree limbs or bridges—may cause reflections that will appear as reflections at depth on the GPR record. The use of shielded antennae is recommended for all surveys; however, if a shielded antenna is not available, an unshielded antenna can be wrapped in heavy butcher paper as a field expedient shield.

Because radar antennae are pulled over the ground surface, rough or bumpy ground may alter GPR data characteristics, such that a ground surface dip may, in fact, be interpreted as a subsurface anomaly. High-frequency antennae (above 300 MHz) are affected more by this than low-frequency antennae.

Law enforcement communications systems utilize frequencies that are very close to those used in GPR surveying. Radio communications originating within 50 ft of a GPR antenna may cause interference with radar signals. It is recommended that radio (walkie-talkie, car radio) communications be kept to a minimum during any GPR surveying.

6.6 Documentation

A GPR record header is also recorded with the GPR data. Most data storage nowadays is on flash media or the Cloud. Information contained in the header includes date, time, time–depth range in nanoseconds, antenna frequency, and dielectric constant. The operator's field notes should include the grid orientation, direction of antenna movement, geologic and soils information, surface conditions, electrical interferences, and the location of any obstacles and cultural features. Because an interpreter will examine each recorded profile, these annotations are extremely important. Because the interpreted information will most likely be transferred to a site map, line location and direction must be accurately recorded.

6.7 Data Reduction

Generally, with the exception of performing computer color enhancement and filtering of data, little data reduction is required. The most obvious data reduction involves conversion of the nanosecond time scale to an approximate depth scale (depth below the ground surface in feet or meters). Because penetration is dependent upon material properties, the knowledge of radar signal propagation in soil materials is helpful. This information can be obtained from the results of GPR surveying over the test area, or from the literature. For example, the two-way travel time of a radar signal in air is 1 ns/ft. Table 6.1 presents

Table 6.1 GPR (EM) Velocity Values of Typical Materials

Material Constant	*Dielectric (ft/10 ns)*	*Depth (ft)*
Soil	16.0	1.25
Gravel, dry	25.0–35.0	0.85–1.00
Sandy soil, dry	2.6	3.10
Sandy soil, wet	25.0	1.00
Loamy soil, dry	2.5	3.16
Loamy soil, wet	19.0	1.14
Clayey soil, dry	2.4	3.22
Clayey soil, wet	15.0	1.29
Sand, dry	4.0–6.0	2.0–2.5
Sand, wet	30.0	0.91
Silt, wet	10.0	1.58
Clay, wet	8.0–12.0	1.44–1.76
Sandy clay, saturated	15.0	1.29
Air	1.0	5.0
Freshwater	81.0	0.55
Freshwater ice	4.0	2.5
Seawater	81.0–88.0	0.53–0.56
Seawater ice	4.0–8.0	1.77–2.50
Glacial ice	3.2	2.8
Permafrost	4.0–5.0	2.2–2.5
Firn (snow)	1.4	4.2
Basalt, wet	8.0	1.76
Shale, wet	7.0	1.88
Limestone, dry	7.0	1.88
Limestone, wet	15.0	1.29
Sandstone, wet	6.0	2.04
Granite	8.0	1.77
Concrete, uncracked, dry	6.0	2.04
Concrete, cracked, dry	4.5	2.35

Source: Modified from Ulriksen, C., Peter, F., Application of impulse radar to civil engineering, Lund University of Technology, Department of Engineering Geology, Doctoral Thesis, 1982.

dielectric constants and corresponding depths, as a function of travel times, for typical materials that might be encountered in forensic investigations. A rough estimate of depth can be determined by utilizing these values.

Radar data can be computer processed to enhance signal and to position reflected signals to their proper location.

6.8 Data Interpretation

Data analysis involves examination of all the graphic profiles, individually and collectively. It is important to recognize that the radar profiles are not a representation of a geologic cross-section. The features observed on a radar profile represent changes in electrical properties that may or may not coincide with subsurface geologic and/or lithologic conditions.

Normally, interpretation involves identifying anomalous features within the graphic profiles. Plotting these features from line to line may allow the interpreter to recognize significant patterns. For example, clandestine graves often exhibit increased water saturation in the subsurface and may be of significantly different composition than the host material. Therefore, this type of feature may be readily apparent on a GPR profile. Overall, the general interpretative approach is to observe patterns that differ from the normal background data.

Computer enhancement processing during the interpretative phase is most useful where subtle features are to be examined. In cases where features are pronounced, additional processing may not be necessary.

6.9 Data Presentation

Radar results are often presented in the forms of 2D profiles (Figure 6.4) and/or 3D maps (Figure 6.5). Where data are acquired in electronic format, profiles can be generated at different scales, different signal enhancement, and/or displayed in color formats. A map of the survey grid, with the site features superimposed, should also be used to show the lateral location and extent of any GPR anomalies. Figure 6.6 represents a GPR anomaly map, with anomalies A, B, and C denoted in shading. The lateral extent of each of these anomalies was detected along each GPR profile (Figure 6.7).

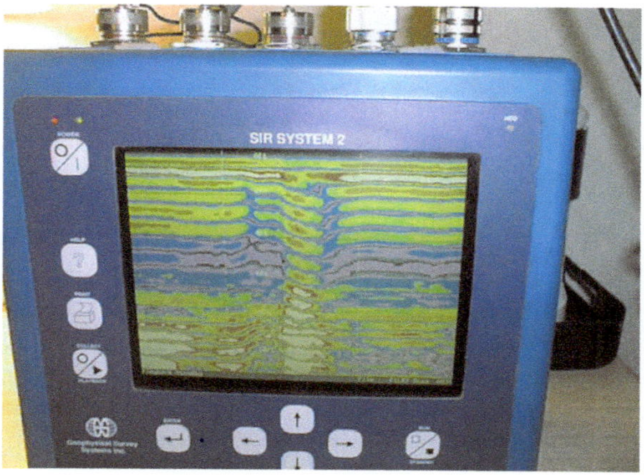

Figure 6.4 GPR profile, on GSSI monitor.

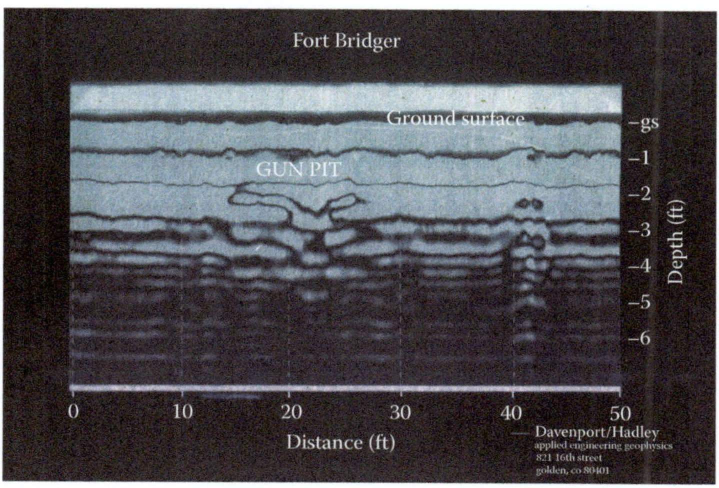

Figure 6.5 GPR 2D radar plot. (From Author's Collection, 1982. Newer plots are shown in color.)

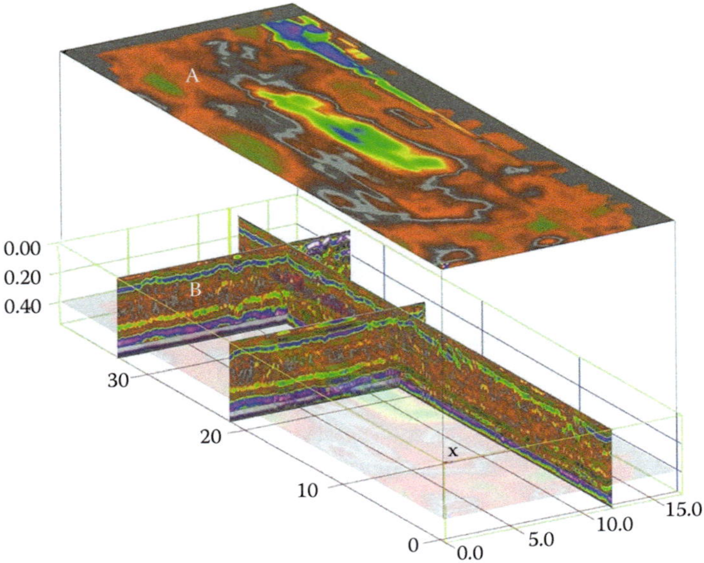

Figure 6.6 (A) 2D GPR plot and (B) 3D GPR cross-section. (Courtesy Rockware.)

6.10 Summary

The author understands the important role that GPR surveys can play in forensic investigations, however, with the following caveats:

- GPR surveys will not detect nor delineate buried human remains
- GPR anomalies will need to be excavated to determine their cause

Finally, the results of a GPR survey were used in a murder conviction in Arizona. Based on this, the author has assisted a number of agencies in wording warrants for the use of GPR.

6.11 Words for a Warrant to Utilize GPR

GPR is often used in forensic investigations. Its operation is based on the introduction of a relatively high-frequency electromagnetic signal into the ground via a moving antenna. As this signal passes through the earth and encounters materials of varying electrical properties, the signal is reflected back to the antenna. This reflected signal is recorded

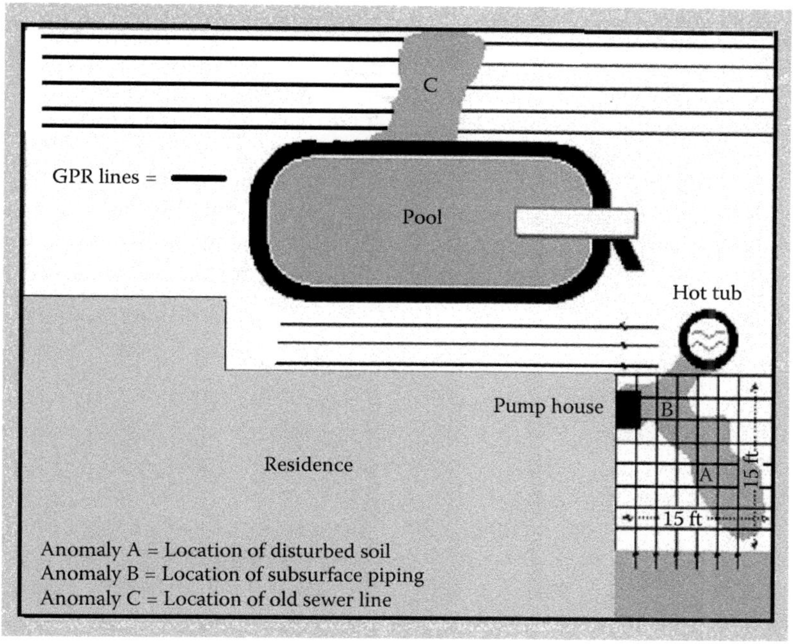

Figure 6.7 GPR survey map. Upon excavation, anomaly A yielded the body of a woman missing 28 years. (From Author's Collection.)

as a color chart depicting a vertical depth cross-section of the earth beneath the antenna.

6.11.1 Advantages

The depth of investigation with GPR is generally quite shallow (<20 ft); however, this disadvantage is partially offset by the increased resolution of the reflected signals, which can show very detailed shallow subsurface conditions.

GPR cannot detect human remains, but it is often an ideal tool to detect and delineate disturbed subsurface materials. In GPR surveying for forensic investigations, disturbed subsurface materials are the principal targets. In addition to detecting and delineating disturbed materials, buried metallic objects are often detected depending on their size and depth.

6.11.2 Disadvantages

The variability of subsurface electrical properties determines the propagation of radar energy through materials. Therefore, adverse subsurface conditions limit its usefulness on a site-specific basis. Water saturation and the chemical nature generally control the dielectric and conductive properties of materials. High-conductivity materials such as saltwater and clay will result in attenuation and a reduction in signal velocity or strength causing a decrease in depth penetration. For instance, the presence of highly conductive clay materials in concentrations of 10% or more is probably the greatest limiting factor affecting the radar method.

6.11.3 Precedents

GPR was successfully used to detect and delineate an anomalous area under concrete that contained the body of a woman missing for 28 years. This work was done, in Arizona in 1994, by G. Clark Davenport, a California registered geophysicist with NecroSearch International. The results of the GPR survey led to the first-degree murder conviction of the woman's husband, Lyle Eugene (Gene) Keidel in April 17, 1995. In April 1997, the Maricopa Court of Appeals upheld Keidel's first-degree murder conviction (Maricopa Superior Court, Criminal Court Cases Number CR1994-008312).

7 Metal Detector Surveying

7.1 Introduction

Numerous varieties of metal detectors, developed primarily for the treasure hunter or for utility locating purposes, are commercially available. Their construction and operating principles are similar to electromagnetic (EM) and magnetic surveying instruments. The ability of metal detectors to quickly locate both ferrous and nonferrous metallic objects at very shallow depths, their relatively simple operation, and their low cost offer the forensic investigator a useful tool.

Metal detectors are useful for identification of metallic objects within a few feet of the surface. EM varieties are generally capable of detecting all conductive metals. They are configured with a radio transmitter–receiver coil arrangement similar to an EM terrain conductivity meter. Magnetic locators are configured in the form of a vertical gradient magnetometer or gradiometer. They are only capable of detecting ferrous objects. Specialized, very sophisticated alkali-vapor detectors are currently used by specialty firms and the military for locating unexploded ordnance.

Utilization of metal detectors in combination with more sophisticated geophysical equipment can provide valuable information to the investigator. Metal detectors can be utilized prior to conducting magnetic, EM, and ground-penetrating radar surveys to delineate for the removal or avoidance of near-surface metallic objects that may produce interference to other types of geophysical surveys. Metal detector surveys can also be used to provide a qualitative idea of the depth of anomalies delineated in magnetic or EM surveys. Because the effective depth of penetration of most metal detectors is known, the absence of a reading with a metal detector when sweeping over a known magnetic or EM anomaly means that the anomaly is deeper than the effective depth of the metal detector.

7.2 Limitations

Metal detectors are generally useful only for investigation of the upper few feet of the surface. Most detectors have an audible tone indicator, either via headphones or an external loudspeaker, to provide the operator with an indication of the presence of a metallic object. Some newer instruments are configured to provide a visible, relative scale of signal amplitude which in turn is correlated to the size and depth of the located object (Figure 7.1). The field of investigation of individual detectors varies, affecting the instrument sensitivity and consequently the discrimination capabilities of object size and depth of burial. In some cases, a small, shallow object may produce an identical instrument response as a large, deeper object. All detectors are subject to signal interference owing to any nearby metallic objects such as fences and underground utilities.

Occasionally, metal detectors of all types may respond to nonmetallic features such as an excavation or dramatic changes in the composition of soils. These features may be detected as a tonal change in the instrument response. The detection of these features can be of

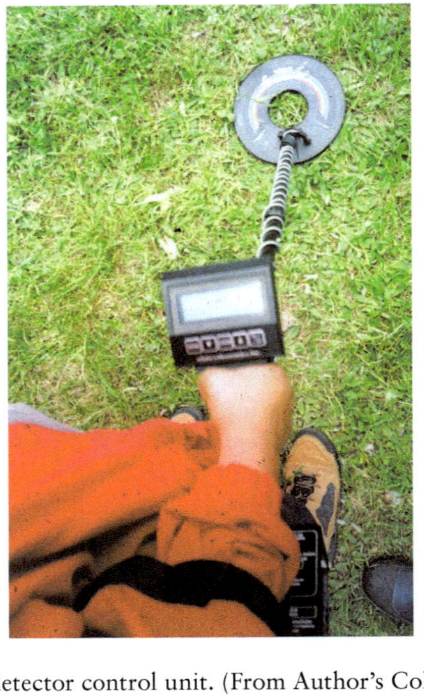

Figure 7.1 Metal detector control unit. (From Author's Collection.)

importance to the investigator looking for a clandestine grave. It may be very difficult or frustrating to use metal detectors in areas of "black sand," such as at a beach; however, many detector manufacturers have built-in filter systems to reduce the effects of black sands. Black sand consists of weathered material containing magnetite, which will cause an almost continuous response (noise) for many detectors.

Metal detector manufacturers use different frequencies to produce their transmitted signals. This can result in interferences, such as false positives, in cases where detectors of different manufacture are used at the same time in an investigation.

7.3 Instrumentation

The detection capabilities of metal detectors are limited by the size, depth, and orientation of the object being searched for.

Three types of detectors are commonly available:

- Coil systems
- Two-box systems
- Magnetic systems

Coil systems consist of a thin, dish-shaped sensor head mounted on a carrying staff. An electronic control box is situated on the opposite end of the carrying staff. The staff is typically 4 ft in length, and may be fitted with a wrist support for operator comfort (Figure 7.2). The sensor head contains concentric coils: a transmitting coil and a receiving coil (Figure 7.3). Many of the coil systems offer interchangeable sensor heads of various diameters. Larger diameter heads are capable of greater depths of investigations, normally in the two-foot range, or for underwater work. Different coil sizes can be used, interchangeably, on some makes of metal detectors. Recent advances in coil systems have been based on military and demining operations and include compact detectors such as the Costruzioni Elettroniche Industriali Automatismi S.p.A. (CEIA) compact metal detector (Figure 7.4).

Two-box systems consist of a three-foot long rod fitted with a transmitting coil on one end and a receiving coil on the other (Figure 7.5), very similar to the configuration of the Geonics EM-38. This system is sling-carried several inches above and level to the ground surface. If the coils are not level with the ground surface, the result can be an erroneous instrument response that may be mistaken for metal. The field of investigation is about 3 ft. Two-box systems have increased depth capabilities over coil systems, but the sensitivity is not as great.

Figure 7.2 Metal detector. (From Author's Collection.)

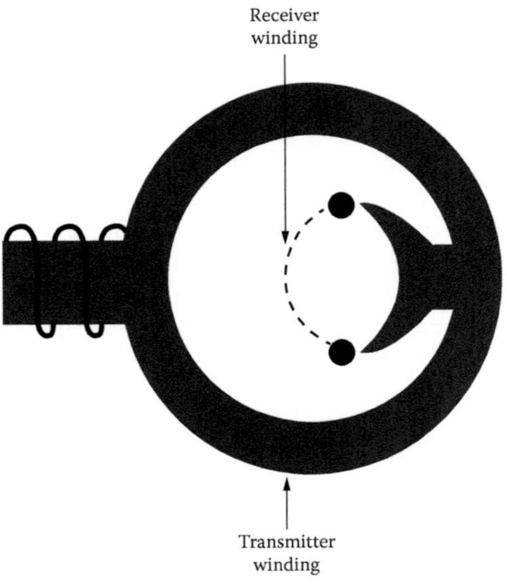

Figure 7.3 Metal detector sensor head. (From Author's Collection.)

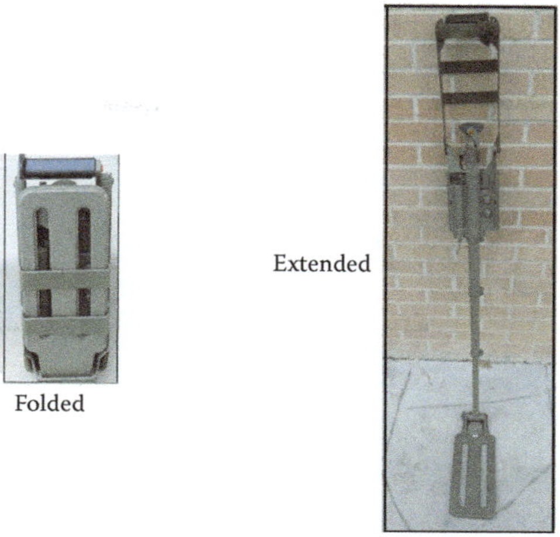

Figure 7.4 CEIA compact metal detector. (From Author's Collection.)

Figure 7.5 Two-box metal detector. (Courtesy of CEIA USA Ltd.)

Magnetic systems, sometimes called valve and box, or pipe and valve locators, are similar in operating principle to magnetic gradiometers. Magnetic detectors are very sensitive to ferrous objects to depths of 8 ft. Partially oxidized (rusted) iron may also cause a response. Two magnetic sensors, separated by 18 in. of nonmagnetic material, are placed inside a nonmagnetic staff (Figure 7.6).

Figure 7.6 Valve and box locator. (From Author's Collection.)

Both sensors will respond to buried iron objects; however, owing to its closer proximity to an object, the lower sensor will produce a greater response than the upper sensor. If the object is small, the upper sensor may not respond to it at all. The sensors produce an audible sound and also, in some instruments, an indication on a meter. With experience, an operator may be able to obtain an idea of the depth, size, and orientation of a detected object based on the signal strength.

7.4 Field Procedures

A metal detector search should be conducted in a systematic manner. Object size, depth of burial, and metallic characteristics will dictate search sweep patterns. A survey grid should be established as in other types of geophysical surveying. The detector operator(s) should slowly walk along each grid line while sweeping from side to side with the detector (Figure 7.7). A full sweep should extend approximately 6–8 ft on either side of the grid line. Upon completion of a full sweep, the operator should advance forward a set distance, normally one pace.

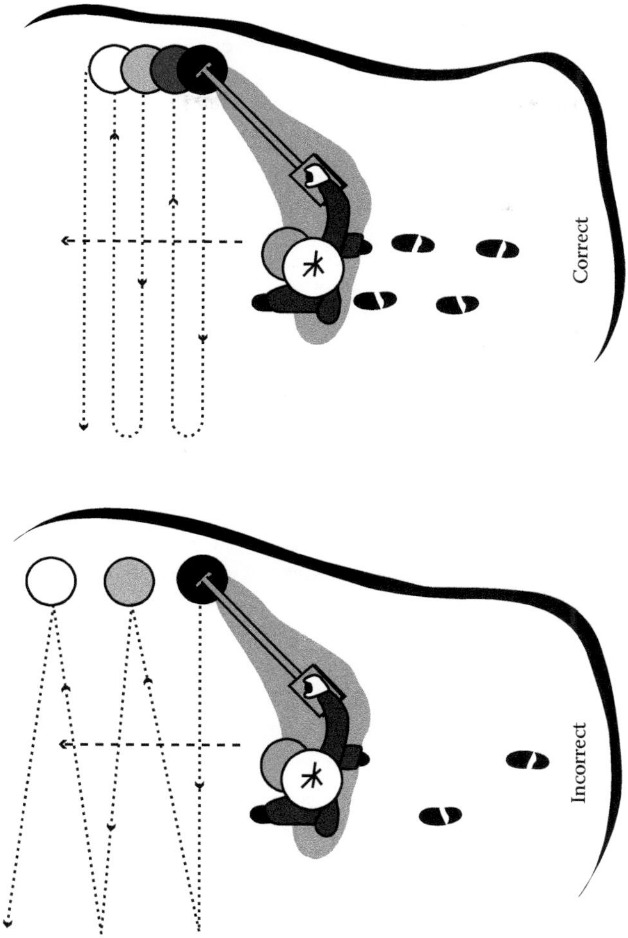

Figure 7.7 Metal detector sweep pattern. (From Author's Collection.)

Every effort should be made to maintain the detector head at a constant height above the ground surface. This may require some subtle adjustments as the head passes over rough terrain.

Each detector operator should be accompanied by an assistant if possible. When a detector response is noted, the operator should slowly and carefully sweep the immediate area of the response in an attempt to pinpoint the target. Experienced operators are able to determine the general size, shape, orientation, and depth of a target. Once pinpointed, the target should be marked with flagging. It is recommended that targets not be excavated as they are encountered. Once the full grid is swept, all flagged targets should be mapped in. After excavation, smaller metal detector search coils can be used to sweep an excavated area to ascertain that the cause of the anomaly has been either removed or remains. The same smaller coils can also be used to sweep any excavated dirt to determine if it contains the cause of any anomaly.

7.5 Special Considerations

The use of metal detectors can be very problematic too. Many agencies have detectors; however, they may no longer have personnel familiar with the detector operations. Any manual or training video that came with the detector may be long lost such that a person selected to use the detector may face a quandary. This situation can often be overcome by making a call to the manufacturer.

When sweeping with a number of operators and detectors, a staggered pattern should be used. Detectors transmit an EM signal with a certain frequency, and different detector manufacturers use different frequencies. Interfering signals may result if different detectors are working closely together. The detector operators may interpret the interfering signals as signs of metal. It is also possible that metal may not be detected owing to interfering signals. Field testing with detectors separated by different distances will help establish the separation distances to maintain between detectors in an actual grid survey.

7.6 Summary

Prior to any metal detector survey, it is recommended that an object similar to the one being searched for, if known, be placed on the ground and "playing" with the detector settings, and the operator sweeps parameters until a response is received.

To avoid false positives caused by closeness of metal detectors of the same or different manufacturer, an optimum separation between detectors should be made prior to the start of any metal detector survey.

The author recommends that any detector "hits" be marked. Once the detector survey is finished, the marked "hits" should be photographed, and then visually examined to determine what patterns, if any, exist. Then the "hits" can be excavated.

Pipe and valve locators can often be obtained for use from local water departments or surveying companies.

8 Infrared

8.1 Introduction

Thermal imaging is based on capturing electromagnetic radiation with wavelengths longer than that those of visible light. Infrared radiation (IR) is emitted by any object having a temperature above absolute 0. The amount of radiation increases as the temperature of an object increases. When imaged, warm objects in a field of cooler objects are readily identified. Thermography, the science of IR, is well suited for forensic investigations.

One of the primary commercial uses of thermography has been mapping structures to obtain thermograms that may indicate heat losses. Firefighters use thermography in situations where heavy smoke obscures the interior of a building. Law enforcement has used thermography as a surveillance tool to track suspects or vehicles at night. Many agencies are equipped with helicopter-mounted surveillance systems, often referred to as forward-looking infrared (FLIR).

NecroSearch International has pioneered the use of thermography for the detection of clandestine graves and evidence. Disturbed ground, such as one would expect for a gravesite, radiates heat differently than the surrounding undisturbed ground, thus providing a heat anomaly when being imaged in the early evening.

8.2 Limitations

The only limitation of thermography is that an object must have a temperature above absolute 0 to radiate IR. When using a camera to capture an IR image, there are a number of techniques that can be used by the photographer to enhance the images. Through one of the leading manufacturers, FLIR systems, Internet training is offered to improve a camera operator's mastery of obtaining optimal thermograms.

Figure 8.1 Thermal imager, FLIR C2. (Courtesy of FLIR Systems.)

8.3 Instrumentation

The resulting images of IR are called thermograms, and these can be recorded by a thermographic camera, such as the FLIR One or FLIR C2 (Figure 8.1). The FLIR C2 is equipped to obtain thermograms and a black and white photograph of the objects being studied. The thermograms obtained from the FLIR C2 can be stored in the camera, and retrieved via a universal serial bus connection, for processing.

8.4 Documentation

Thermograms obtained on the FLIR C2 are documented on a header for each image. This documentation includes image number, time, and date.

8.5 Data Processing

Processed FLIR C2 thermograms (Figure 8.2) can provide measurements between –10°C and 150°C at any point on an image, although this level of detail may not be necessary in forensic investigations.

Figure 8.2 Thermal image, Russian Cathedral. (From Author's Collection.)

FLIR systems have software that can be utilized to process data taken by the FLIR C2.

8.6 False-Color Infrared

False color refers to an image that shows an object in a color that differs from its natural color. For example, a healthy green bush would appear in a false-color image as a red bush, hence the term false color (Figure 8.3). This effect is the result of recording an image using a different electromagnetic wavelength than is used for thermal infrared images. False-color images are recorded by using special film. This type of film requires special processing. Newer digital cameras can provide, through the use of infrared filters, ready images.

False-color images are useful to the criminal investigator by pointing out contrasts between healthy vegetation, imaged as red, and unhealthy vegetation, imaged as brown. The location of an area of unhealthy vegetation may indicate disturbed soil.

Figure 8.3 Natural color image: (A) false-color infrared image and (B) Necro-Search research site. *Note*: In (B) the red color indicates healthy vegetation, while the brown area, left center, indicates disturbed vegetation (grave sites of three pigs.) (From Author's Collection.)

8.7 Summary

Thermal imaging can play a very important role in the forensic investigations by assisting investigators in locating disturbed ground. The imaging must be taken when a heat contrast exists, typically in the early evenings.

Thermal imaging equipment, if not owned by an agency, can most likely be obtained from local fire department and/or military units.

In lieu of having a thermal camera available, the author has used a noncontact digital thermometer (Figure 8.4) to register heat anomalies. In particular, a basement is heated up to a maximum using an industrial grade heater (e.g., similar to those used by football teams during the cold games). The floor and walls of the basement are previously gridded off in 2×2 or 3×3 foot square boxes using masking tape.

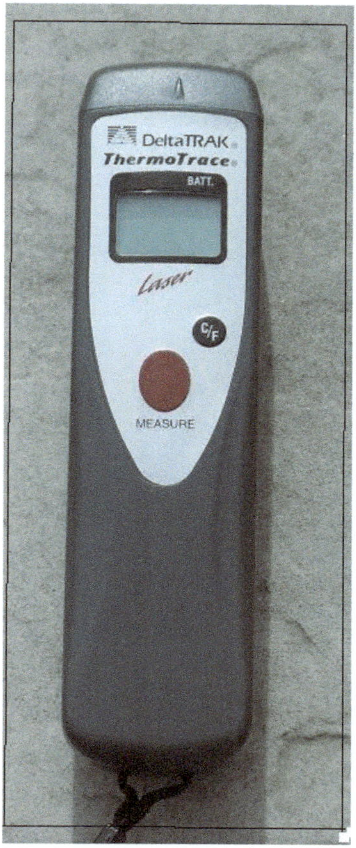

Figure 8.4 Noncontact digital thermometer. (From Author's Collection.)

Once maximum heat is obtained, the heater is shut down, and as the basement cools the temperature of each grid square is taken periodically. This crude system allows an investigator to look for and map heat anomalies that may be caused by disturbances.

In the event that a false-color infrared imaging camera is not available, polarized sun or shooting glass can be used as a substitute for viewing an area. The results will not be as dramatic or as high resolution as that of a false-color imaging camera but will help the investigator in locating areas of disturbed vegetation. The author uses polarized amber lens glasses for sunny days, and polarized purple lens glasses for overcast or hazy days.

9 Marine Geophysical Methods

Although waterborne (marine) environmental geophysical surveys conducted for forensic investigations are rare, a basic understanding of marine geophysical techniques may be valuable to investigators. Several geophysical methods can effectively be used to investigate submerged sites. Acoustic surveying and magnetic surveying have been utilized to discover and delineate numerous shipwreck and archaeological sites in both fresh- and saltwater environments. Ground-penetrating radar (GPR), useful in freshwater but not salty or brackish waters, has been used experimentally in shallow freshwater environments to investigate scouring around bridge piers.

Loran navigational and global positioning systems (GPS) are used for location and positioning during large-scale marine geophysical surveys, whereas conventional land surveying techniques are used for positioning small-scale surveys.

9.1 Acoustic Methods

Useful acoustic methods consist of side-scan sonar, tuned transducers, fathometers, and sub-bottom profilers.

Side-scan sonar (Figure 9.1) emits and receives reflected high-frequency acoustic pulses in the 50–500 kilohertz (kHz) range from targets across the seafloor. It produces a very high-resolution image of the water bottom (seafloor or lake floor). In a sense, side-scan sonar is to marine investigations as aerial photography is to terrestrial investigations. Side-scan sonar does not penetrate into the water bottom subsurface. In operation, an instrument package, called a fish, is towed, at a predetermined depth, behind a support vessel. The depth of the fish is dependent on the sub-bottom topography, the desired field of view, and the water conditions.

Figure 9.1 Side-scan sonar. (From Author's Collection.)

Sub-bottom profilers and fathometer systems (such as fish finders) operate very similarly to side-scan sonar, with respect to a source and receiver. They are different in that they use lower frequency energy transmissions, in the 5–50 kHz range, and a downward signal beaming to reflect off and penetrate the seabed. In operation, a sub-bottom profiler is towed, at a relatively shallow depth, behind the support ship, while a fathometer is mounted on the hull of the ship. For lake bed searches, smaller remote control sub-bottom systems are available, or a fish can be towed behind an inflatable boat. Sub-bottom profiling methods are performed with lower frequency acoustic transducers and acoustic sources called sparkers or boomers. The lower frequency signal is capable of penetrating the bottom sediments to depths of up to 50 ft. These systems are primarily utilized in seafloor hazard studies and other engineering applications.

9.2 Magnetic Methods

Marine magnetic surveying is performed in a systematic manner similar to a land survey. Because perturbations in the total geomagnetic field are sought, similar anomalies should be observable in a water environment. Most marine surveys are conducted with a two-sensor gradiometer system to avoid making diurnal corrections. The magnetometer is towed by a ship. Because ships are generally magnetic, the sensor is towed hundreds of meters behind the ship. Magnetic surveys are used primarily to detect shipwrecks, active submarines, and bottom or sub-bottom communications cables. The use of marine magnetometers in forensic investigations is rare (The author has used a valve and box locator to search for a bicycle in about 10 ft of water.)

9.3 Ground-Penetrating Radar

GPR can be used in shallow freshwater approximately 50 ft deep. Because freshwater generally exhibits a low conductivity, radar energy will penetrate freshwater to various depths, depending on the antenna frequency. Also, depending on the electrical characteristics of the bottom sediments, sub-bottom penetration may also be possible, giving GPR a distinct advantage over fathometers, which offer very little sub-bottom penetration. The survey profiling is performed by placing the antenna in an inflatable raft (or nonmetallic boat) and pulling or towing the raft across the water on a grid system (Figure 9.2). Position navigation is

Figure 9.2 GPR survey on freshwater. (From Author's Collection.)

always a problem when performing GPR surveys in water. It may be beneficial to perform GPR surveying on ice, provided a water body be subject to winter freezing.

Recent innovative applications of radar in freshwater include bridge scour surveys, locating submerged materials, determining sedimentation rates behind dams, and counting fish.

9.4 Special Considerations

Many waterborne geophysical surveys, such as magnetics and GPR, can be run using an ice surface as the surveying platform. This means that if an agency is unable to obtain marine geophysical equipment for surveying in open water, it may be advisable to wait until winter when the water freezes over, and then run standard land-based geophysical surveys.

10 Airborne Geophysical Methods

Magnetic and electromagnetic instrumentation can be used for airborne surveying, from fixed- or rotary-wing platforms. The instrument packages, termed birds, are suspended from or flown below the platforms. The surveys are typically flown along flight lines separated by fixed distances determined by the target parameters. Data are continuously recorded in digital format. Navigational control can be obtained via global positioning systems (GPS) and/or other systems such as minirangers and video cameras.

During surveying, all attempts are made to keep the instrument packages at constant heights above the ground surface. Surveys are flown in either a contour or drape mode. Contour flying involves keeping the instrument package at a constant altitude, while drape flying involves keeping the instrument package at a constant elevation above the ground surface. Both elevation and altitude are monitored by laser altimeters to permit the pilot to maintain the desired altitude. Instrument packages also contain downward-looking video cameras to provide a true visual presentation of the ground surface directly below the flight lines.

Airborne geophysical surveying has little practical application in forensic investigations, because target sizes are extremely small in comparison to flight heights. Airborne geophysical surveys have, however, been successfully applied in the detection and delineation of clandestine hazardous waste sites and as such can play a role in the investigation of environmental crimes.

11 Quality Control

Geophysical surveying can play an important role in criminal investigations. Should the results of a survey be useful in the apprehension of a suspect, the geophysical data, and survey methodology may be subject to scrutiny by the courts. For this reason, the geophysicist must strive to obtain the best quality data, using industry-accepted survey techniques and calibrated equipment.

To assure the quality of the data, it is extremely important that properly trained personnel (i.e., geophysicists) be involved in the planning, data acquisition, analysis, and interpretation elements of each geophysical survey. The report covering the results of any survey should discuss quality assurance procedures utilized.

11.1 Records

The investigating geophysicist will keep field and administrative records according to his or her experience. Depending on the anticipated legal proceedings that may result from the criminal investigation, the following records may be useful.

11.1.1 Field Activity Records

It is recommended that a daily diary of field activities performed during the investigation be kept. This may include descriptions of changes and/or modifications to the geophysical surveys, listings of personnel and visitors to the scene, and a general assessment of survey performance and progression.

11.1.2 Equipment Use Log

A log of geophysical equipment used should be maintained. Information concerning equipment operating hours, serial numbers, calibrations performed, known deficiencies, and associated corrective actions are normal reporting requirements.

11.2 Original Data Sets

Experienced criminal investigators are very familiar with time, effort, and expense required to conduct field operations. Geophysical operations are performed under similar circumstances. Because much of the acquired geophysical field data is unique, it must be considered irreplaceable, meaning that it must be archived and stored with care.

Transmittal of any data should be done through accepted Chain-of-Evidence (Chain-of-Custody) procedures. This is a system that requires a responsible party to maintain custody of evidence at all times. In the event that another party wishes to use or examine the evidence, that party must sign a Chain-of-Custody form acknowledging receipt of the evidence. This procedure is designed to keep valuable evidence items in control at all times, so that they cannot be tampered with.

11.3 Data Storage

Upon survey completion, either the geophysicist or the criminal investigator will store the data. All hard copy and electronic data should be appropriately labeled, catalogued, entered into Chain-of-Custody, and stored carefully.

In the event that the survey data are used in judicial proceedings, and entered into evidence, it must be maintained in storage until all appeals, if any, are exhausted.

12 Concluding Remarks

For an optimal outcome, and one which will withstand professional and legal scrutiny, and for results that can be admitted as evidence in a court of law, the requirements for an ideal survey would consist of the following:

1. The criminal investigator: communicating an absolutely clear understanding of survey objectives and acting as the sole point of contact for the geophysicist
2. Management: arranging sufficient logistical support, to assure freedom from interference by supervisors and media personnel
3. The geophysicist: have a thorough understanding of survey objectives, site constraints, crime scene procedures, and law enforcement requirements
4. The site: low-noise, high-contrast targets and controlled access
5. Data acquisition: the ability to acquire data utilizing all applicable methods
6. Time: sufficient allocation of time to reduce and interpret all data
7. Ground truth: prioritization of all pertinent geophysical anomalies for future excavation

Table 12.1 presents a summary of the applicability of geophysical techniques for forensic investigations. The presentation of these applicable forensic investigative methods will hopefully be helpful to criminal investigators.

Understanding the capabilities, limitations, and logistics of geophysical surveying can only be realized by experiencing the frustrations and joys of utilizing these techniques to assist in finding the proverbial needle in the haystack.

Table 12.1 Forensic Applications of Geophysical Techniques

Item/Technique	*Magnetics*	*Electro-Magnetics*	*Metal Detector*	*Ground-Penetrating Radar*	*Infrared (IR)*	*Side Scan Sonar*	*Comments*
Shell casings	Not useful	Not useful	Useful	Not useful	Possible	Not useful	
Bullets	Not useful	Not useful	Useful	Not useful	Possible	Not useful	
Unexploded ordnance	Useful	Useful	Useful	Possible	Condition dependent	Condition dependent	IR for surface material only
Shallow grave	Useful	Useful	Possible	Condition dependent	Condition dependent	Not useful	
Grave under concrete/asphalt	Not useful	Not useful	Not useful	Useful	Possible	Not useful	
55 gal drum 5–10 ft deep	Useful	Useful	Not useful	Useful	Not useful	Not useful	
Body under water	Not useful	Not useful	Not useful	Possible	Not useful	Useful	
Metal property in water	Useful	Useful	Useful	Possible	Not useful	Useful	
Nonmetal property in water	Not useful	Not useful	Not useful	Condition dependent	Not useful	Useful	

Not useful; Useful; Possible; Condition dependent.

Glossary

A

Amplitude The size of a signal, either in the ground or after amplification. Usually measured from the 0 or rest position to a maximum excursion.

Anomaly A deviation from uniformity in physical properties, often of interest in mineral/oil exploration. An area or feature of contrasting geophysical response.

Apparent conductivity One of the quantities measured during an electromagnetic induction survey. It is proportional to the actual conductivities of subsurface materials.

Apparent resistivity The electrical property measured during a resistivity survey. It is proportional to the actual resistivities of subsurface materials.

Apparent velocity (1) The velocity with which a wave front registers on a line of geophones. (2) The inverse slope of a time–distance curve.

Archaeogeophysics Prospection in European terms. The use of geophysical methods to evaluate an archaeological site.

B

Bedrock Solid rock exposed at the surface of the earth or overlain by unconsolidated material.

C

Character The recognizable aspect of a geophysical response, usually in the waveform, which distinguishes it from other events. Usually a frequency or phasing effect, often not defined precisely and hence dependent upon subjective judgment.

Conductivity The ability of a material to transfer an electric current. It is equal to the inverse of resistivity.

Conductivity meter (also called ground conductivity meter or terrain conductivity meter) An instrument used to measure the strength of the electromagnetic field generated within the earth by an induced current.

Current The quantity of charge transmitted per unit time.

Cross-section A plot of geophysical or geological events. A seismic cross-section, GPR cross-section, time versus distance, or depth versus distance plots.

D

Datum (1) The arbitrary reference level to which measurements are corrected. (2) The surface from which seismic reflection times or depths are measured, corrections having been made for local topographic and/or weathering variations. (3) The reference level for elevation measurements, often sea level.

Declination The lateral attitude of the earth's magnetic field at any location on the earth's surface.

Dielectric The capacity of a material to store a charge in the presence of an electrical field.

Digital Representation of quantities in discrete units. An analog system is one in which the information is represented as a continuous flow of the quantity constituting the signal. A digital system is one in which the analog data is sampled at discrete intervals for later reconstruction into analog form.

Diurnal drift Drift in the earth's magnetic field due to external sources such as solar flares and solar wind storms.

E

Electromagnetic Periodically varying electromagnetic fields, such as light, radio waves, and radar transmissions.

Electromagnetic (EM) survey A geophysical exploration method whereby electromagnetic fields are induced in the ground and the resultant secondary magnetic field is detected and interpreted in terms of ground conductivity.

F

Ferrous Any metal that can be magnetized. Any metal that will rust (i.e., iron, steel).

Filter (1) The part of a system which discriminates against some of the information entering it. The discrimination is usually on the basis of frequency, although other bases such as wavelength or moveout (see velocity filter) may be used. The act of filtering is called convolution. (2) Filters may be characterized by their impulse response or more usually by their amplitude and phase response as a function of frequency. (3) Bandpass filters are component filters. (4) Notch filters reject sharply at a particular frequency. Primarily used to reject 60-Hz line noise. (5) Digital filters provide a means of filtering data numerically in the time domain by summing weighted samples at a series of successive time increments.

Frequency The repetition rate of a periodic waveform measured in cycles per second (cps) or Hertz (Hz). Angular frequency measured in radians per second.

G

Gamma Unit of magnetic field intensity.

Geophone (Seismometer) Instrument used to convert seismic energy into electrical energy.

Gravimeter Gravity meter, an instrument for measuring variations in gravitational attraction.

Gravity survey A survey performed to measure the gravitational field, or derivatives, to associate differences in the field with density distribution and therefore differences in rock types.

H

Hydrophone A pressure detector sensitive to variations in pressure, as opposed to a geophone which is sensitive to motion. Used when the detector can be placed below a few feet of water.

I

Inclination The dip of the earth's magnetic field at any location on the earth's surface. (A measure of metallic concentration).

Inphase Electrical signal with the same phase angle as that of a transmitted signal.

Isoconductivity (iso = equal) Map of lines of equal conductivity.

L

Line A series of stations connected together.

M

Magnetic survey A geophysical survey method that depends on detection of susceptibility contrasts (anomalies) caused by the presence of material that can be magnetized.

Magnetic susceptibility Property of a material corresponding to its ability to distort an applied magnetic field.

Magnetometer A device used for precise and sensitive measurements of magnetic fields.

Magnetometry The science of measuring variations in the earth's magnetic field.

Millimho Unit of conductance, reciprocal of ohm, (mmho).

Millisecond Measurement of time used in acoustic or seismic surveying (one thousandth of a second).

Millivolts Unit of electrical measurement used in self-potential (SP) surveying.

N

Nanosecond Measurement of time used in GPR surveying (one millionth of a second).

Nanotesla Unit of measurement of the intensity of the earth's magnetic field (also gamma).

Noise (1) Any undesired signal; a disturbance which does not represent any part of a message from a specified source. (2) Sometimes restricted to energy which is random. (3) Seismic energy which is not resolvable as reflections. In this sense, noise includes microseisms, shot-generated noise, tape-modulation noise, harmonic distortions, etc. Sometimes divided into coherent noise (including nonreflection coherent events) and random noise (including wind noise, instrument noise, and all other energy which is noncoherent). To the extent that noise is random, it can be attenuated by a factor of n by compositing n signals from independent measurements. (4) Sometimes restricted to seismic energy not derived from the shot explosion. (5) Disturbances in observed data due to more or fewer random changes in surface and near-surface material.

O

Original data Any element of data generated directly in the field in the investigation of a site, or a new element of data resulting from a direct manipulation or compilation of the field data.

P

pH Measurement of acidity or alkalinity.
Potential The voltage with respect to a reference point.
Profile The series of measurements made from several locations from which a cross-section or can be constructed.

Q

Quadrature phase Electrical signal 90° out of phase with the transmitted signal. (A measure of conductivity).

R

Radar A system in which short electromagnetic waves are transmitted into a medium, and the energy which is scattered back by reflecting objects is detected. May be used for shallow penetration surveys in the ground or over ice.
Reconnaissance (1) A general examination of a region to determine its main features, usually preliminary to a more detailed survey. (2) A survey whose objective is to ascertain regional geological structures or to determine whether economically prospective features exist, rather than to map an individual structure.
Resistivity (electrical) A property of rock material giving a measure of the difficulty involved in driving an electrical current through it. Mathematically, resistivity is the ratio of electric field intensity to current density.
Resistivity meter A general term for an instrument used to measure in situ the resistivity of soil and rock materials.
Resistivity survey A survey performed to observe the electric fields and earth resistivity caused by introducing a current into the ground.

S

Seismic survey A survey performed to determine geologic and structural conditions within the earth. An acoustic signal is input into the ground, reflecting or refracting from subsurface layers of different physical properties. The reflected or refracted signals are recorded on the ground surface by geophones.
Seismic velocity The rate of propagation of seismic wave through a medium.

Self-potential Spontaneous potential, natural potential, SP. The dc or slowly varying natural ground voltage between nearby, nonpolarizing electrodes.

Signal enhancement A hardware development utilized in seismographs and resistivity systems to improve signal-to-noise ratio by real-time adding (stacking) successive waveforms from the same source point and thereby discriminating against random noise.

Silicification The process by which a material is hardened by the addition of silica or silaceous materials and heat.

Site data package Accumulation of geophysical data information from a single source of dissemination analysis and evaluation against the site criteria. Would include charts, records, field notes, maps, photographs, technical reports, lab reports, etc.

T

Telluric Natural electrical earth currents of low amplitude covering wide regions.

Trace A record of one seismic channel. This channel may contain one or more geophones. A trace is made by a galvanometer.

V

Vibroseis A seismic energy source consisting of controlled frequency input into the earth by way of large vibrators (truck mounted).

W

Wavelength The distance between successive similar points on two adjacent cycles of a wave, measured perpendicular to the wave front.

WWV The U.S. Bureau of Standards radio station that broadcasts time and frequency standards.

For the definition of other geophysical terms used in this manual, refer to: *Glossary of Terms Used in Geophysical Exploration*, Society of Exploration Geophysicists, Tulsa, Oklahoma, 1984.

Selected References

General Works

Bevan, B. (1991). The search for graves. *Geophysics*, 56, 1310–1319.

Bevan, B., & Weymouth, J. (1982). A compendium of recent geophysical surveys. *Society for Archaeological Sciences Newsletter*, 6, 3–4.

Bevan, B., & Weymouth, J. (1984). Recent geophysical surveys of archaeological sites. *Society for Archaeological Sciences Newsletter*, 7, 4–6.

Bray, D.E., & Stanley, R.K. (1989). *Nondestructive Evaluation*. New York: McGraw-Hill.

Brothwell, D., & Higgs, E. (1962). A test of magnetometry as an aid to archaeology. *American Antiquity*, 28, 199–205.

Clark, A. (1975). Archaeological prospecting: A progress report. *Journal of Archaeological Science*, 2, 297–314.

Clark, A. (1990). *Seeing Beneath the Soil*. London: Batsford.

Davenport, G.C. (1989). Forensic geophysics. *Proceedings of the Le Cochon Connection, Search Methods and Techniques for Locating Clandestine Graves, Rocky Mountain Division of the International Association for Identification, Spring Conference*, Englewood, CO, 1, 1–12.

Davenport, G.C. (2001). Remote sensing applications in forensic investigations. *Historical Archaeology*, 35, 87–100.

Davenport, G.C. (2001). *Where Is It? Searching for Buried Bodies and Hidden Evidence*. Church Hill, MD: SportWork.

Davenport, G.C., Griffin, T.J., Lindemann, J.W., & Borowski, J. (1988). Geotechnical applications, crime scene investigation techniques. *Geophysics, The Leading Edge of Exploration*, 7, 64–66.

Davenport, G.C., Griffin, T.J., Lindemann, J.W., & Heimmer, D. (1990). Geoscientists and law officers work together in Colorado. *Geotimes*, 35, 13–15.

Dobrin, M.B., & Savit, C.H. (1988). *Introduction to Geophysical Prospecting*. New York: McGraw-Hill.

Ebert, J.I. (1984). Remote sensing applications in archaeology. *Advances in Archaeological Method and Theory*. New York: Academic Press.

France, D., Griffin, T.J., Swanburg, J., Lindemann, J.W., Davenport, G.C., Trammel, V., Armburst, C.T., Kondratieff, B., Nelson, A., Castellano, K., & Hopkins, D. (1992). A multidisciplinary approach to the detection of clandestine graves. *Journal of Forensic Sciences*, 37, 1445–1458.

Hochrein, M.J. (2001). *A Bibliography Related to Geotaphonomic and Forensic Archaeological Field Techniques*. Saint Louis, MO: US Department of Justice, Federal Bureau of Investigation.

Hoving, G.L. (1986). Buried body search technology. *Identification News*, 3, 15.

Imaizumi, M. (1974). Locating buried bodies. *FBI Law Enforcement Bulletin*, 2, 2–5.

Killam, E.W. (1990). *The Detection of Human Remains*. Springfield, IL: Charles C. Thomas.

Leith, C.J., Schneider, K.A., & Carr, C. (1976). Geophysical investigation of archaeological sites. *Bulletin of the International Association of Engineering Geology*, 14, 123–128.

McManamon, F.P. (1984). Discovering sites unseen. *Advances in Archaeological Method and Theory*. New York: Academic Press.

Northwest Mining Association. (1989). *Practical Geophysics for the Exploration Geologist*. Spokane, WA.

Parrington, M. (1979). Geophysical and aerial prospecting techniques at Valley Forge National Historic Park, Pennsylvania. *Journal of Field Archaeology*, 6, 193–201.

Ward, S.H. (1990). *Geotechnical and Environmental Geophysics, Vol. 1, Review and Tutorial*. Tulsa, OK: Society of Exploration Geophysicists.

Wynn, J.C. (1986). Review of geophysical methods used in archaeology. *Geoarchaeology*, 3, 245–257.

Magnetics

Barrows, L., & Rocchio, J.E. (1990). Magnetic surveying for buried metallic objects. *Ground Water Monitoring Review*, 10, 204–209.

Breiner, S. (1973). *Applications Manual for Portable Magnetometers*. San Jose, CA: Geometrics.

Breiner, S., & Coe, M.D. (1972). Magnetic exploration of the Olmec civilization. *American Scientist*, 60, 566–575.

Johnston, R.B. (1961). Archaeological application of the proton magnetometer in Indiana (USA). *Archaeometry*, 4, 71–72.

Ralph, E.K. (1965). Comparison of a proton and a rubidium magnetometer for archaeological prospecting. *Archaeometry*, 7, 20–27.

Tabbagh, A. (1984). On the comparison between magnetic and electromagnetic prospection methods for magnetic features detection. *Archaeometry*, 26, 171–182.

Weymouth, J.W., & Bevan, B. (1983). Combined Magnetic and Ground Penetrating Radar Survey of an Archaeological Site on Oklahoma. San Francisco, CA: IEEE Digest, I, Section WP.3.

Electromagnetics

Bevan, B. (1983). Electromagnetics for mapping buried earth features. *Journal of Field Archaeology*, 10, 47–54.

McNeill, J.D. (1980). Electromagnetic Terrain Conductivity Measurement at Low Induction Numbers. Technical Note TN 6. Mississauga, ON: Geonics Limited.

McNeill, J.D. (1985a). *Operating Manual for EM31-D Non-contacting Terrain Conductivity Meter.* Mississauga, ON: Geonics Limited.

McNeill, J.D. (1985b). *Operating Manual for the EM-38 Terrain Conductivity Meter.* Mississauga, ON: Geonics Limited.

Parchas, C., & Tabbagh, A. (1979). Simultaneous measurements of electrical conductivity and magnetic susceptibility of the ground in electromagnetic prospecting. *Proceedings of the 18th International Symposium on Archaeometry and Archaeological Prospection*, 10, 682–691.

Tite, M.S., & Mullins, C. (1970). Electromagnetic prospecting on archaeological sites using a soil conductivity meter. *Archaeometry*, 12, 97–104.

Ground-Penetrating Radar

Bevan, B., Orr, D.G., & Blades, B.S. (1984). The discovery of the Taylor house at the Petersburg National Battlefield. *Historical Archaeology*, 18, 64–74.

Daniels, J. (1988). Locating caves, tunnels and mines. *Geophysics: The Leading Edge of Exploration*. Tulsa, OK: Society of Exploration Geophysicists.

Davis, J.L., & Annan, A.P. (1989). Ground penetrating radar for high-resolution mapping of soil and rock stratigraphy. *Geophysical Prospecting*, 37, 531–551.

De Vore, S.L. (1990). Ground penetrating radar as a tool in archaeological investigations: An example from Fort Laramie National Historic Site. *The Wyoming Archaeologist*, 33, 23–28.

De Vore, S.L. (1991). *Subsurface Interface Radar 10 Manual.* New Salem, NH: Geophysical Survey Systems, Inc.

Heimmer, D.H., Davenport, G.C., Lindemann, J.W., & Gilmore, J.B. (1989). Geophysics for archaeological assessment: Ft. William discovered? Ft. Laramie National Historical Site, Wyoming. *Proceedings of the Symposium on the Application of Geophysics to Engineering and Environmental Problems*, 448–455.

Imai, T., Sakayama, T., & Kanemori, T. (1987). Use of ground-probing radar and resistivity surveys for archaeological investigations. *Geophysics*, 52, 137–150.

Mellett, J., & Geismar, J. (1990). GPR survey of an African-American cemetery in Little Ferry, NJ. *Proceedings of the Third International Conference on Ground Penetrating Radar*, Lakewood, CO, 42.

Olhoeft, G.R. (1988). Selected bibliography on ground penetrating radar. *Proceedings of the Symposium on the Application of Geophysics to Engineering and Environmental Problems*, Golden, CO, 462–520.

Persons, S. (1990). Burial detection using ground penetrating radar. *Proceedings of the Third International Conference on Ground Penetrating Radar*, Lakewood, CO, 54.

Ulriksen, C.P. (1983). *Application of Impulse Radar to Civil Engineering*. New Salem, NH: Geophysical Survey Systems.

Metal Detection

Garrett, C. (1985). *Modern Metal Detectors*. Dallas, TX: Ram Publishing Company.

Graham, R.K. (1995). Metal detection. *Law Enforcement Bulletin*, 64, 10–15.

Fischer, P.M. (1980). Geophysical prospecting at Hala Sultan Tekke, Cyprus. *Journal of Field Archaeology*, 7, 479–484.

Fischer, P.M. (1986). *Instruction Manual Model MAC-51B Magnetic and Cable Locator*. Reston, VA: Schonstedt Instrument Company.

Index

Note: Page numbers followed by *f* indicate figures; those followed by *t* indicate tables.